Thinking Mathematically

We work with leading authors to develop the strongest educational materials in mathematics, bringing cutting-edge thinking and best learning practice to a global market.

Under a range of well-known imprints, including Prentice Hall, we craft high quality print and electronic publications which help readers to understand and apply their content, whether studying or at work.

To find out more about the complete range of our publishing, please visit us on the World Wide Web at:
www.pearsoned.co.uk

Thinking Mathematically

Second Edition

John Mason
University of Oxford/Open University

with Leone Burton
and
Kaye Stacey
Melbourne Graduate School of Education
University of Melbourne

Prentice Hall
is an imprint of

Harlow, England • London • New York • Boston • San Francisco • Toronto
Sydney • Tokyo • Singapore • Hong Kong • Seoul • Taipei • New Delhi
Cape Town • Madrid • Mexico City • Amsterdam • Munich • Paris • Milan

Pearson Education Limited
Edinburgh Gate
Harlow
Essex CM20 2JE
England

and Associated Companies throughout the world

Visit us on the World Wide Web at:
www.pearsoned.co.uk

First published 1982
Second edition published 2010

ISBN: 978-0-273-72891-7

British Library Cataloguing-in-Publication Data
A catalogue record for this book is available from the British Library

Library of Congress Cataloging-in-Publication Data
A catalog record for this book is available from the Library of Congress

10 9 8 7 6 5 4 3 2 1
14 13 12 11 10

Typeset in 9.5/13 pt Avenir Book by 73.
Printed and bound in Great Britain by Henry Ling Ltd, Dorchester, Dorset

Contents

Introduction to First Edition

Thinking Mathematically is about mathematical processes, and not about any particular branch of mathematics. Our aim is to show how to make a start on any question, how to attack it effectively and how to learn from the experience. Time and effort spent studying these processes of enquiry are wisely invested because doing so can bring you closer to realizing your full potential for mathematical thinking.

Experience in working with students of all ages has convinced us that mathematical thinking can be improved by

- tackling questions conscientiously;
- reflecting on this experience;
- linking feelings with action;
- studying the process of resolving problems; and
- noticing how what you learn fits in with your own experience.

Consequently while encouraging you to tackle questions, we show you how to reflect on that experience by drawing your attention to important features of the process of thinking mathematically.

How to use this book effectively!

Thinking Mathematically is a book to be used rather than read, so its value depends on how energetically the reader works through the questions posed throughout the text. Their purpose is to provide recent, vivid experience which will connect with the comments that are made. Failure to tackle the questions seriously will render the comments meaningless and empty, and it will be hard to use our advice when it is needed. Three kinds of involvement are required: physical, emotional and intellectual.

Probably the single most important lesson to be learned is that being stuck is an honourable state and an essential part of improving thinking. However, to get the most out of being stuck, it is not enough to think for a few minutes and then read on. Take time to ponder the question, and continue reading only when you are convinced that you have tried all possible alleys. Time taken to ponder the question and to try several approaches is time well spent. Each question is followed by suggestions under the heading STUCK? to provide signposts when progress seems blocked. Because different resolutions follow different paths, some of the suggestions might be mutually contradictory, or

irrelevant to your approach, so do not expect every suggestion to provide immediate insight!

Recalcitrant questions which resist resolution should not be permitted to produce disappointment. A great deal more can be learned from an unsuccessful attempt than from a question which is quickly resolved, provided you think about it earnestly, make use of techniques suggested in the book, and reflect on what you have done. Answers are irrelevant to the main purpose of this book. The important thing is to experience the processes being discussed.

To stress our concentration on processes rather than answers, a 'solution' in the usual sense is rarely given. Instead, we offer sample 'resolutions' which include a good deal of commentary as well as many false starts, partially digested ideas and so on. Elegant solutions such as are found in most mathematics texts rarely spring fully formed from someone's brain. They are more often arrived at after a long and tortuous period of thinking and not thinking, with much modification and changing of understanding along the way, but most beginners do not realize this. By taking our informal approach, confidence can grow and progress made. Elegance can come later.

In summary, then, our approach rests on five important assumptions:

1 **You** can think mathematically.
2 Mathematical thinking can be **improved** by practice with reflection.
3 Mathematical thinking is **provoked** by contradiction, tension and surprise.

4 Mathematical thinking is **supported** by an atmosphere of questioning, challenging and reflecting.

5 Mathematical thinking helps in **understanding** yourself and the world.

You will notice that the text is written in the first person singular despite there being three authors. This reflects our way of working, as well as the fusion that has taken place during writing.

This book is addressed to students as a manual for developing mathematical thinking. It presents only one approach to the task and does not, for example, compare that approach with the schemes of earlier writers such as Pólya. A bibliography is provided for the reader who wishes to inspect personally the work which has most strongly influenced us.

Whilst some of the problems used in this book are original, many have come from the mathematical grapevine. We would like to thank the friends and colleagues who have introduced us to these questions and to thank especially their usually unknown authors for the enjoyment they have given us.

We are particularly indebted to:

George Pólya and J. G. Bennett for their inspiration;

Graham Read for the cartoons of PIX who first appeared in *Mathematics: A Psychological Perspective*, Open University Press, 1978;

Alan Schoenfeld for stressing the importance of the Monitor discussed in Chapter 7;

Mike Beetham for help in processing the text on the Cambridge computer;

numerous colleagues, most especially Joy Davis, Susie Groves, Peter Stacey and Collette Tasse; and

a myriad of students in three countries.

We offer the book as support to future thinking, particularly of

Quentin and Lydia Mason,
Mark Burton,
Carol and Andrew Stacey.

Introduction to Second Edition

Thinking Mathematically was published in 1982 and continues to find favour in many different countries. It is used by senior high school students, students going on to study mathematics at university, teacher preparation courses, and in courses for undergraduates in mathematics. Our aim in this new edition is to offer a range of questions for exploration appropriate to readers as pre-service primary and secondary teachers, and as mathematics undergraduates. These can be found in the new Chapter 11. Whereas the questions (problems) in the original book were chosen to illustrate the various 'processes', or, as we would now put it, the use of various natural powers, the questions in Chapter 11 have been chosen to make use of those powers to enrich and deepen appreciation of core ideas of various important mathematical topics.

A by-product is a demonstration of the way in which ordinary questions, designed to be approached routinely, can sometimes be transformed into intriguing questions. It also demonstrates that significant areas of higher mathematics and sometimes difficult mathematical questions often lie hidden just behind elementary topics. At the same time, we take the opportunity to rephrase the language of *thinking processes* used in the original book, into the language of *natural powers* which all human beings possess. This also provides an opportunity to include some insights and distinctions that have emerged in the period since the original was published.

Processes and natural powers

In the 1970s and early 1980s, there was great interest in the 'processes' by which things were done, and thinking mathematically is a fine example. However, while interest in delineating processes of thinking and creativity have become of renewed interest recently, the language in which they are described has changed considerably. We found that it made more sense to us, and to people with whom we engaged mathematically and pedagogically, to think in terms of natural powers that learners bring to the classroom. The task of teaching then becomes one of provoking learners to make use of and to develop those powers in the context of mathematical thinking.

We follow Caleb Gattegno in seeing awareness as the basis for action; without awareness, there is no action. However, some awarenesses may be so integrated into our functioning that we are not consciously aware of them operating. This is certainly the case when we suddenly find ourselves acting automatically out of habit. Again following Gattegno, mathematics as a discipline only arises when people become aware of the actions they are performing in certain contexts (relationships and properties in number and space) and articulate these awarenesses to produce 'mathematics'. So mathematics as a body of knowledge in books can be seen as formal recognition, expression and study of awarenesses that inform mathematical actions in problematic situations. To become a teacher requires becoming aware of the awarenesses that generate mathematical actions, because these are what trigger pedagogical actions. Consequently it is vital to educate one's awareness by engaging oneself in mathematical tasks which bring important mathematical awarenesses to the surface, so that they can inform future action.

Awareness is closely related to cognition; action is closely related to behaviour. An often overlooked aspect of human psyche is the emotions or the domain of affect. The original book addressed this through suggesting that being stuck is an 'honourable state' from which it is possible to learn, and that expressing emotion-laden observations about being stuck and having an insight (AHA!), however transitory, releases energy which enables progress to be made. It celebrated positive emotions: the pleasure of making sense through use of your own powers, the excitement of discovery, the aesthetic pleasure in an interesting result, and the satisfaction of finding a resolution. We add here that developing a disposition to recognize problematic situations in the material world as well as in the world of mathematics, the 'questioning attitude' of Chapter 8, is also a significant contribution to the affective domain.

The current emphasis on collaborative activity as a necessary component of mathematics learning is a development from the value recognized and promulgated through the original book, that working together can be stimulating and can open up avenues that no single individual might have recognized by themselves. At the same time, it is vital to have periods of 'own thinking' during which possibilities are considered and either pursued or dropped. Some people like to start individually, and then, after a period, exchange possibilities; others like to have a period of collective idea-generation followed by own thinking before coming together again. Certainly it is helpful to have communal reflection as a force to bring to the surface and articulate insights and observations about salient moments in the exploration, even though these will often have occurred during individual thinking. The presence of significant others is an effective contribution to stimulating the impulse to express and so clarify your own thinking, as well as to connect it to the thinking of others.

We also take the opportunity afforded by this new edition to introduce persistent and ubiquitous mathematical themes which imbue mathematics. A brief description of powers, themes and related notions can be found in the new Chapter 12.

The power of an experiential approach

The original book was conceived as an exposition of our own experience as mathematical thinkers, profoundly influenced by the work of George Pólya. Indeed, JohnM had been shown his film *Let Us Teach Guessing* in 1967 as a graduate teaching assistant soon after it was made, and it released in him an approach to teaching which he later realized was moulded by his experience in high school, where he had been taught by Geoff Steele. To his surprise he discovered many years later that Geoff had never trained as a teacher, and was not primarily a mathematician, but rather a choir conductor! Nevertheless, his stimulation nurtured and sustained John through high school and into university where he arrived having internalized the elements of mathematical thinking.

On arriving at his first academic post at the Open University, John discovered that one of Pólya's books had been chosen as a set book. When he was asked to design a one-week summer school for up to 7,000 students over 11 weeks on three sites, he incorporated the film into the programme, accompanied by sessions called *active problem solving*. John naively assumed that all mathematics tutors would 'be mathematical with and in front of their students' and so would naturally get students specializing and generalizing, conjecturing and convincing and so on. It took some years before he realized that not all tutors were as self-aware of their own mathematical thinking as he had assumed. The result was a series of training sessions for tutors, designed to get them to experience mathematical thinking for themselves and to reflect on that experience so as to be able to draw student attention to important aspects. Meanwhile, the course had been redesigned and so the summer school programme was modified accordingly, with more stress on simpler problems which nevertheless highlighted a specific 'process' of mathematical thinking, or, put another way, which provoked learners to make spontaneous use of one or more natural powers which are important in thinking mathematically.

In 1979 John became involved in his first mathematics education in-service course with Leone, who had been teaching primary school teachers to work mathematically with their students, and researching the effects of this on children's learning. The course team wanted the course to be practical, so the experiential basis was extended by assigning a period of study each week to what was called 'own thinking'. The idea was that, in order to be alert and sensitized to students, it is necessary to be alerted to correspondingly relevant aspects of your own thinking. The problem was how to choose problems and commentary to put into the 'own thinking' sections. In order to make a sensible choice, Leone and John planned the book and were later joined by Kaye who from the other side of the world had also been inspired by Pólya's expositions of mathematical discovery and had for several years been fostering this for primary and secondary pre-service teachers through her innovative mathematical problem-solving courses run jointly with Susie Groves. *Thinking Mathematically* made use of one of the principles being proposed in the course, namely that doing and talking are vital activities to prepare for recording, and that recording helps to integrate doing and talking so as to make insights and experience available to inform action in the future. In our case, the writing of the book crystallized and organized our own thinking about what experiences would be most useful for teachers.

As authors, we attribute the continuing interest in and use of the book to its experiential basis, and that continues in the present edition. Indeed an intention of the new edition is to make it easier for teachers to bring the experience of thinking mathematically more centrally into all teaching. Developing your mathematical thinking, indeed engaging in discussion about any mathematics education issue, is greatly improved by engaging together in related mathematics first, and then looking for other shareable experiences on which to draw. Put another way, all of the great educational theorists who have addressed mathematics education agree that learning is enhanced when students are given tasks which spark off activity in which familiar actions are adapted and modified in order to meet the challenge. There is little use in rehearsing problems you can already do, using familiar actions, unless you are simply trying to gain speed. Activity produces experience, but as Immanuel Kant might have said,

> **A succession of experiences does not add up to an experience of that succession.**

Something more is required. Pólya called it the stage of *looking back*. We chose to call it the Review phase, which is more precise than the term *reflecting* which we also use. *Reflection* has a multiplicity of meanings. Jim Wilson once said that this stage was the most talked about and the least used of Pólya's four stages. Most educationalists are agreed that some sort of drawing back from immersion in activity is necessary in order to learn from experience. After all,

> **One thing we do not seem to learn from experience is that we do not often learn from experience alone.**

Where educationalists differ is in the timing, the degree and the initiative of drawing back from the action that is required. Clearly drawing back too soon leaves everyone frustrated and is unlikely to have a lasting effect. On the other hand, leaving students to 'learn from their experience' is clearly unsatisfactory for all but the most gifted of mathematics students. For most students, learning to learn mathematics is a scientific endeavour rather than a natural endeavour, in the sense of Lev Vygotsky: most people need to be in the presence of someone more experienced, at least some of the time, in order to make sense of experience. Caleb Gattegno and others argue that learning really takes place during sleep, when the mind chooses what to forget, or at least to let go of. Be that as it may, where students have been engaged in practices of intentional reflection, review, reconstruction and rehearsal, they are much more likely to have access to insights in the future. The Discipline of Noticing which JohnM articulated based on his experience with J. G. Bennett was an attempt to provide a philosophically well-founded method for researching one's own practice, but it applies equally well to students learning.

In order to learn from experience, in order to have fresh possibilities come to mind when appropriate, it is necessary to sensitize yourself to notice opportunities to respond to situations rather than to react, to choose to act rather than to be caught up in and driven by old habits. Thus, by offering tasks and follow-up prompts it is possible to promote awareness of the use of people's natural powers. As they become more sensitized to and aware of their own use of these powers, those powers develop in flexibility and in usefulness. They become, in the language of Vygotsky 'actions for oneself', which can be self-initiated, rather than simply 'actions in oneself', which have to be triggered by a teacher or cued by some prompt in the task. All this provides some justification for the format both of the previous book and of this one: questions are provided to be worked on. They are followed by reflective prompts and commentary. They are of little or no use unless they are engaged in fully, perhaps over a considerable period of time, followed by reflection and looking out for resonance between comments provided and own experience. Our aim is to promote pondering and contemplating, getting stuck and restarting. Getting 'answers' is not the most valuable outcome of struggling. Rather, what is most important and valuable is what you notice happening in the way of getting stuck, making progress, making conjectures, modifying conjectures, using your powers, encountering mathematical themes, etc., and the little frissons of insight and excitement that come from using those powers and making some progress. The tasks are fodder for initiating activity; the mathematical results are not usually of significance. Put another way, this book does not try to teach any particular mathematical content, but rather to alert readers to the ways in which their own natural powers can be harnessed in the service of exploring and understanding mathematical topics and situations.

There is always an issue of level of challenge. Initial impressions of a question may lead to a sense of 'too challenging' or of 'not challenging enough'. One of the things to learn from working on questions is how to make something less challenging so that progress can be made (usually by specializing) and more challenging, by locating some *dimensions of possible variation* and varying them, or by extending the *range of permissible change* of those features. It is up to the reader to select the level of challenge appropriate to them at any given moment, with the hope that they will be inspired to return and tackle the more challenging tasks at a later date. The intention is that the questions posed will initiate mathematical experience and that readers and their teachers will adapt the degree of difficulty to make this experience productive, rather than focusing exclusively on getting answers to fixed problems.

Acknowledgements

When the original book was written, we believed that mathematical questions belonged in the world of mathematics, without the need to provide details of their origins. Age has brought with it interest in the origins of problems and how they are transformed over time. Where we believe we know of a specific origin, we have inserted that in this new edition. Where no reference is given, either its origins have been forgotten or it came to us from several different colleagues as it passed from person to person through the community of mathematical thinkers, or we believe that we have constructed it ourselves. We are grateful to comments on the new additions by Eva Knoll and Ami Mamolo.

Dedication

The original book was dedicated to our children, who of course have now grown up. Sadly Leone lost her fight against cancer before we got started on these revisions, so we offer this edition in her memory. In the words of her son, Mark:

> Leone Burton's books were always dedicated to me, her son. This book is dedicated to her, and her memory. Certainly problem solving, and thinking mathematically, have been the richest gift I have had; whether it came from her, or indeed another author of this book, John Mason, who first brought a computer to our house when I was a very young boy, so that I could 'play Logo'.

John Mason, Oxford, April 2010
Kaye Stacey, Melbourne, April 2010

1

Everyone can start

This chapter introduces the activities which will get your thinking started on any question. There is no need to shy away from a mathematical question, and no reason to stare at a blank piece of paper feeling hopeless. Driving straight down the first path that appears hoping brute force will succeed is not a good tactic either. However, there are productive things you can do.

Specializing

The best place to begin is to work on a question:

Warehouse

In a warehouse you obtain 20% discount but you must pay a 15% sales tax. Which would you prefer to have calculated first: discount or tax?

How can you get to grips with such a question? To make progress, you must be clear what the question is asking, but this may not fully emerge until you've done a bit of doodling. The best way to start is by trying some specific cases. I hope you spontaneously want to try it with an item priced at say £100.

DO SO NOW IF YOU HAVE NOT ALREADY

Surprised by the result? Most people are, and it is that surprise which fuels mathematical thinking. Now, will the same thing happen for a price of say £120?.

TRY IT AND SEE!

Write down your calculations and your insights. It is the only way to develop your thinking skills.

TRY IT AND SEE!

Now, perhaps using a calculator, try other examples. Your aim in doing this is two-fold: to get an idea of what the answer to the question might be, and at the same time to develop a sense of why your answer might be correct. Put another way, by doing examples you make the question meaningful to yourself and you may also begin to see an underlying pattern in all the special cases which will be the clue to resolving the question completely.

What might be the underlying pattern in this question? Perhaps you have experience of questions like this and know what to do. If so, think how you would encourage someone less experienced to tackle it, then read my suggestions. It is important to work through my discussions because that is where important points about mathematical thinking will be introduced and illustrated.

How does the final price depend on the order of calculating discount and tax? There should be a pattern in the examples you have tried. If not, check your calculations! Will this result be true for other prices? If you are not certain, try some more examples. When you are sure, search for an explanation (or read further).

TRY EXAMPLES UNTIL YOU ARE SURE

A lot depends on the form in which you do your calculations. The usual form for doing discount followed by tax is to

calculate the discount:	on £100 discount is £20
subtract it from the price:	£100 − £20 = £80
calculate the tax:	15% of £80 is £12
add the tax on to get the	
final price:	£80 + £12 = £92

Try to find other ways of doing the calculation until you hit upon one which reveals why your result is always true. As a suggestion, you want to find a form of calculation which is independent of the initial price. To do this, try calculating what percentage of an original price you pay when the discount has been subtracted, and what percentage of an original price you pay when tax has been added.

DO IT NOW

With any luck you will have found that

(i) subtracting 20% from a price is the same as paying 80% of it, that is you pay 0.80 times the price:

(ii) adding 15% to a price is the same as paying 115% of it, that is you pay 1.15 times the price.

Then, for any initial price of say £100, calculating

discount first: you pay $1.15 \times (0.80 \times £100)$

tax first: you pay $0.80 \times (1.15 \times £100)$

By writing the calculation in this form you can see that the order of calculation does not matter, because all that is involved is multiplying the original price by two numbers, in either order. If the original price is £P then calculating

discount first: you pay $1.15 \times 0.80 \times £P$

tax first: you pay $0.80 \times 1.15 \times £P$

and these are always equal.

Notice the value of standing back from the detail of the calculation and looking at its form or shape. This sort of reflective activity is fundamental to developing your mathematical thinking.

Warehouse illustrates several important aspects of mathematical thinking, two of which I want to draw to your attention. Firstly there are specific processes which aid mathematical thinking. In this case the process being emphasized is SPECIALIZING which means turning to examples to learn about the question. The examples you choose are special in the sense  that they are particular instances of a more general situation in the question. Secondly, being STUCK is a natural state of affairs, and something can usually be done about it. Here, the something being suggested is SPECIALIZING. This is a simple technique which everyone can use, and when people find themselves unable to proceed with a question, suggestions like

Have you tried an example?

and

What happens in this particular case?

are what gets them going again.

The next question, taken from Banwell, Saunders, and Tahta (1986), illustrates other forms of specializing.

Paper Strip

Imagine a long thin strip of paper stretched out in front of you, left to right. Imagine taking the ends in your hands and placing the right hand end on top of the left. Now press the strip flat so that it is folded in half and has a crease. Repeat the whole operation on the new strip two more times. How many creases are there? How many creases will there be if the operation is repeated 10 times in total?

TRY IT NOW

STUCK?

➤ Specialize mentally by counting the creases after two folds.
➤ Perhaps a diagram will steady your mental image.
➤ Specialize by trying it on a strip of paper.
➤ Try three folds and four folds. Look for a pattern.
➤ What do you want to find? Be clear and precise.
➤ Is there something related to the creases that you can count more easily?
➤ Check any conjectures on new examples!

I am not going to give a full resolution of this question. If you are STUCK, do not be upset. Being stuck is fine, as long as you look on it as an opportunity to learn. Perhaps you can return to the question with renewed vigour as you read the next chapter! Before you set it aside, try up to five folds either mentally, with diagrams, or with real paper. Count the creases and draw up a table of the results. Whereas with *Warehouse* specializing means turning to numerical examples to get to grips with the question, specializing for *Paper Strip* means turning to diagrams or pieces of paper and experimenting. It is important to turn to objects which you are confidently able to manipulate. These may be physical objects, or mathematical ones such as diagrams, numbers or algebraic symbols.

Specializing alone is unlikely to resolve a question for you, but it does get you started and involved. The question loses its forbidding exterior and becomes less intimidating. Furthermore, the specific cases should help you to get a sense of what the question is really about, enabling you to make an informed guess. Further careful specializing with an eye on the 'why' rather than the 'what' may lead to insight into what is really happening.

The next question is on more familiar ground.

Palindromes

A number like 12321 is called a palindrome because it reads the same backwards as forwards. A friend of mine claims that all palindromes with four digits are exactly divisible by 11. Are they?

TRY IT NOW

STUCK?

➤ Find some palindromes with four digits.
➤ Do you believe my friend?
➤ What do you want to show?

A resolution

Remember that a resolution is not intended to be polished, and is only one way of thinking about it. The only sensible way to begin is to specialize. I want to get a feel for the kinds of numbers involved. What are some palindromes?

> 747 is one
> 88 and 6 are others

The question only mentions palindromes with four digits, which means numbers like

> 1221, 3003, 6996 and 7557.

What do I want? I want to find out if all such numbers are divisible by 11.

TEST IT NOW

By trying specific numerical examples, I convinced myself that the result seems plausible. Notice however that I cannot be sure my result is **always** correct just by specializing unless I am prepared to test **every** four-digit palindrome. As there are some 90 of them, it is better to try to get some idea of the underlying pattern.

DO SO NOW

I tried four specific cases:

> 1221/11 = 111
> 3003/11 = 273
> 6996/11 = 636
> 7557/11 = 687

but I could see no obvious pattern in them. This brings up an extremely important point about specializing. Choosing examples randomly is a good way of getting an idea of what is involved in a question and seeing if a statement or guess is likely to be true, but when searching for patterns, success is more likely if the specializing is done systematically. How can I be systematic in this case?

TRY IT NOW

STUCK?

➤ What is the smallest four-digit palindrome?
➤ What is the next smallest?
➤ How can one palindrome be changed into another one?

One way is to start with the smallest four-digit palindrome (that is 1001) and work upwards in numerical order:

> 1001, 1111, 1221, 1331, . . .

Checking my friend's statement:

$$1001/11 = 91$$
$$1111/11 = 101$$
$$1221/11 = 111$$
$$1331/11 = 121$$

This not only supports my friend's claim, but also suggests more. Notice that the palindromes are rising by 110 each time, and the quotients are rising by 10 each time.

AHA! Now I can see why my friend's claim is true. The difference between successive palindromes is always 110. The smallest palindrome (1001) is exactly divisible by 11 and so is 110. As all the other palindromes are obtained from 1001 by adding on 110, all the palindromes with four digits must be exactly divisible by 11.

So apart from tidying up and expressing it nicely, the question is resolved.

Or is it? Does the resolution cover all the specific cases I have used? Look more closely! If all palindromes can be constructed by successively adding 110 to 1001, they will all have one as the units digit. But they do not! For example, 7557 is a palindrome with 7 as its units digit. What has gone wrong? Specializing led to a pattern (that successive palindromes differ by 110) on which I based my resolution. But this pattern cannot hold for all palindromes because it predicts something that is false (all palindromes do not end in one). The fault lies in jumping too quickly from the three differences to a general result. Fortunately, specializing can help again; this time to pinpoint the weakness in the pattern. Look further along the list of palindromes:

Palindromes	1881		1991		2002		2112		2222		2332
Differences		110		11		110		110		110	

This time I will proceed more cautiously, perhaps in a mood of disbelief rather than of belief. The pattern seems to be that successive palindromes differ by 110 except when the thousands digit changes and then the difference is 11. Further specializing gives results in agreement with this and increases confidence that this is indeed the underlying pattern. Thus specializing has again provided insight into what pattern might be valid. Now it is time to seek a general reason why the new pattern is valid, finally arriving at something like this:

> Successive palindromes that have the same thousands digit must have the same units in order to be palindromes. Thus the numbers differ only in the second and third digits which are each greater by one. The difference is therefore 110.
>
> Successive palindromes that differ in the thousands digit arise by adding 1001 (to increase the thousands and unit digits) and subtracting 990 (to reduce the second and third digits from nines to zeros). But $1001 - 990 = 11$, as observed in the examples.

In both cases, the differences are divisible by 11, so as long as the smallest four digit palindrome (1001) is divisible by 11 (it is), all of them are.

Now look back at the ways in which specializing has been used:

- It helped me to understand the question by forcing me to clarify the idea of a palindrome.
- It also led me to discover the form of a four-digit palindrome.
- I used it to convince myself that what my friend claimed was indeed likely to be true.
- Later on, systematic specializing exposed a pattern and so gave me an idea of why the result was true.
- Testing whether that pattern was correct (it was not) involved further specializing.

It is because it can be used so effectively, so easily, and in so many ways that specializing is basic to mathematical thinking.

The argument given in my resolution is by no means the most elegant, but then my aim is not for elegance in the first instance. The first attempt is rarely like the solutions printed in text books. If you are more mathematically sophisticated and confident with letters standing for arbitrary numbers, then you may easily have reached a resolution more quickly. You may for example have noticed that every four-digit palindrome has the shape *ABBA* where *A* and *B* are digits. Such a number has the value

$$1000A + 100B + 10B + A = (1000 + 1)A + (100 + 1)B$$
$$= 1001A + 110B$$
$$= 11 \times 91A + 11 \times 10B$$
$$= 11(91A + 10B)$$

(If you find this symbolic argument hard to follow, specialize and follow it through with $A = 3$ and $B = 4$. Then use other values for A and B until you have a sense of the patterns being expressed by the symbols.)

Elegant resolutions like that one apparently show no evidence of specializing since, by means of the symbols, a general argument applying to all four-digit palindromes is given. However, in order to create this argument, I must be sufficiently familiar with the entities involved (namely four-digit palindromes, *A*s and *B*s and decimal notation) that the general form *ABBA* is concrete and confidence inspiring. I must be at ease manipulating both the palindromes and the symbols standing for them. This is the essence of specializing. Turning to familiar, confidence inspiring entities and using them to explore what the question is about creates feelings of confidence and ease in otherwise unfamiliar situations.

Generalizing

In the discussion of specializing, it was impossible to avoid the other side of the coin, the process of generalizing: moving from a few instances to making guesses about a wide class of cases.

Generalizations are the life-blood of mathematics. Whereas specific results may in themselves be useful, the characteristically mathematical result is the general one. For example, knowing what happens for an article priced at £100 in *Warehouse* is less powerful than knowing that the final price is always independent of the order of calculation of discount and tax.

Generalizing starts when you sense an underlying pattern, even if you cannot articulate it. After the *Warehouse* calculations had been carried out for a few prices I noticed that, in each case, the order of calculation did not affect the result. This is the underlying pattern, the generalization. I conjectured that the order of calculation would never alter the result. When the calculation was put into a helpful form it was easy to introduce the symbol P for the original price and thereby show that the generalization was true.

Generalizing need not stop here. What if the discount and tax rates change? Does the order of calculation sometimes make a difference?

IF YOU HAVE NOT ALREADY DONE SO, TRY IT NOW

I hope you can see from the shape of the calculation derived earlier that the actual percentages are irrelevant to the argument. Part of the power of symbols in mathematics is to express such a general pattern. In this case denote the discount rate as a decimal or fraction by D, denote the tax rate as a decimal or fraction by V, and denote the original price by P. Then with

discount first: you pay $P(1 - D)(1 + V)$

tax first: you pay $P(1 + V)(1 - D)$

These are always equal because the order in which we multiply numbers (and hence symbols representing numbers) does not change the outcome. The use of symbols enables the argument to be presented concisely and whole classes of examples (in this case, all possible prices, tax rates and discount rates) can be treated at once. However, exploiting symbols is by no means as straightforward as is popularly imagined – it depends on the symbols becoming as familiar and meaningful as the numbers they replace.

Warehouse illustrates in a simple form the constant interplay between specializing and generalizing that makes up a large part of mathematical thinking. Specializing is used to gather the evidence upon which a generalization is to be made. Articulating the pattern that has been sensed produces a conjecture (a shrewd or informed guess) which further specializing can support or demolish. The process of justifying the conjecture involves more generalizing, with a shift in emphasis from guessing what may be true to seeing why it may be true. In *Warehouse* I first generalized the result by conjecturing that changing the order of calculation does not alter the final price (the 'what'). To justify this I had to study the method of calculation (the 'why').

Palindromes illustrates two other important aspects of generalizing. Being systematic in specializing is often an important aid to generalizing because pattern is more likely to be evident among related examples than with randomly chosen ones. There is, however, an inherent danger. Whilst a pattern may stick out, it is easy to be misled into believing the pattern is right when it is too simple and only partly correct. In *Palindromes*, the difference of 11 between some successive palindromes was overlooked because no examples had been tried where the thousands digit changed. Being cautious about believing an observed pattern or generalization reminds you to test it with a variety of examples. This is the bread and butter of mathematical thinking. Being trigger-happy with conjectures is as dangerous as being reticent to guess. The sometimes delicate balance that has to be struck between being too willing to believe a generalization and too sceptical to make any leaps into darkness is discussed in Chapters 5 and 6.

Writing yourself notes

Before we go on to look at more examples of specializing and generalizing, I wish to introduce a technique for recording mathematical experience. The reason for introducing it now is that you should start recording your experiences so that they are not lost, but can be analysed and studied later. Recording your experiences will also help you to notice them and this contributes to developing your mathematical thinking. Aim to record three things:

- all the significant ideas that occur to you as you search for a resolution to a question;
- what you are trying to do;
- your feelings about it.

Obviously this is a tall order, but it is well worth attempting. In particular, it gives you something to do when you get stuck – write down STUCK! Recognizing that you are stuck is the first step towards getting out of it.

Writing down the feelings you have and the mathematical ideas that occur to you will destroy the stark whiteness of the piece of paper that confronts you as you begin a question.

Once a start has been made, ideas often begin to flow more freely. Then it is important to write down what you are trying to do as it is easy to lose track of your approach or the reasons for embarking on some long calculation. There is nothing worse than surfacing from a bit of work and having no idea what you are doing or why!

I suggest that you get into the habit of writing notes to yourself when working on any of the questions in this book. **Do not be put off** by the large variety of things to note down. As the chapters progress I shall be making suggestions about the most useful things to record. The best time to begin is now, so try making notes as you work on the next question. **Avoid describing** what you do. Brief notes which help you recall the moment are all that is needed. Remember to specialize and generalize, and compare your account with mine only when you have done all you can. My account is necessarily more formal than yours will be, and for later reference I have put certain words in capitals.

Patchwork

Take a square and draw a straight line right across it. Draw several more lines in any arrangement so that the lines all cross the square, and the square is divided into several regions. The task is to colour the regions in such a way that adjacent regions are never coloured the same. (Regions having only one point in common are not considered adjacent.) How few different colours are needed to colour any such arrangement?

TRY IT NOW. NOTE DOWN IDEAS AND FEELINGS, RESORTING TO MY COMMENTS ONLY WHEN YOU GET STUCK

STUCK?
➤ Clarify the question by specializing – try colouring an arrangement.
➤ What do you KNOW? How is an arrangement constructed?

➤ What do you WANT to find?
➤ Be systematic!

A resolution

What is the question asking? Try an example (specialize) to see what is going on:

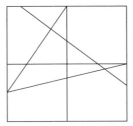

These five lines produce 13 regions. I KNOW that I have to colour the regions so that adjacent ones are different colours. Here is one way using four colours:

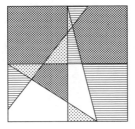

I WANT to find the minimum number of colours needed for any arrangement of lines. Is four the minimum needed for this particular arrangement? TRY to use only three colours.

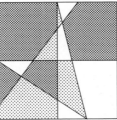

Success! Try again using only two colours.

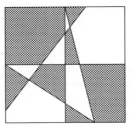

Success again! Obviously one colour is inadequate, so for this particular arrangement, two colours is the minimum required.

As I was filling in the colours, I noticed that I always had to colour 'opposite' regions the same colour (generalizing!).

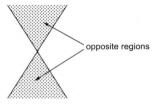

opposite regions

Will two colours always be enough? CHECK with another example – aim for two colours and use the 'opposite' rule (specializing again!).

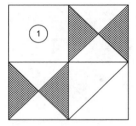

AHA! The 'opposite' rule does not work. When the dark regions have been coloured using the 'opposite' rule, region (1) cannot be dark and yet it cannot be white. Other regions have the same difficulty. Either I need more than two colours or I must abandon the 'opposite' rule. Which route should I follow?

TRY colouring it again with two colours, but abandon the 'opposite' rule.

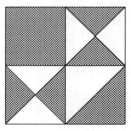

As I made this successful attempt I noticed that once one region is coloured it is easy to work out the rest. The regions adjacent to a coloured region should immediately be allocated the other colour – an 'adjacent' rule. The 'opposite'

rule fails, but now I conjecture that every arrangement of regions can be coloured using only two colours (generalizing to find WHAT might be true).

I do not really have much evidence for this conjecture at the moment. STUCK! How can I convince myself that it always works? AHA! Specialize systematically.

One line:

two colours are enough

Two lines:

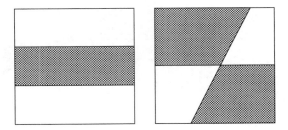

two colours are enough

Three lines:

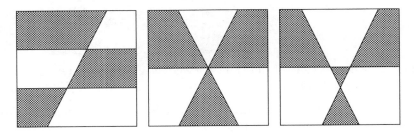

AHA! As I do this systematically, looking at what happens as I add the new line, I begin to see why two colours always suffice (generalizing to find WHY). When I add a new line (say the third) some of the old regions get cut into two parts.

I now keep all the regions (whole and parts of cut regions) on one side of the new line the same colour as they were. On the other side of the new line, I must change the colours of all the regions. See how this works with three lines:

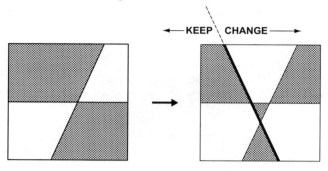

CHECK again: test the method by trying to build up the colouring of the first example (more specializing).

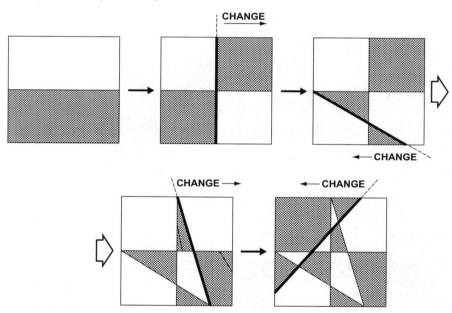

It does work for this example and I think it will always work. The whole square is coloured properly because

1 each side of the new line is properly coloured because adjacent regions were coloured differently by the old colouring;

2 regions which are adjacent along the new line are also coloured differently.

Thus the whole square with the new line added is properly coloured.

Was it just luck that my new method gave the same colouring of the square as my first laborious attempts? What would happen if I added the lines in a different order using the new method? Would this give a different colouring? How many different colourings are possible in this example, and in general? What if the straight lines become curves? What about planes in space? I urge you to look at some of these questions, because it is only by setting a resolution in a broader context that you really come to understand it fully.

DO SO NOW

The other important technique for really understanding and appreciating a resolution is to take time to review what you have done. The notes that you make as you go along are invaluable for this, because it is astonishing how little people remember of what they were doing. It is **not** useful to spend time reconstructing what you 'must' have done. It **is** extremely valuable to review what you **did** do. You might like to compare your notes with mine. You will of course have written rather less than I did and in fact mine have been expanded from my original ones. However, every capitalized word appeared, with shorthanded forms of the reasoning. You will probably have chosen different examples. As you tried colouring, you probably found some sort of pattern or rule, like my 'opposite' or 'adjacent' rule developing inside you. The key thing is to notice it and try to bring it to the surface, capturing it in words. When it is written down it can be examined critically in a way that is not possible when it remains indefinite in your head. As it happened, my 'opposite' rule was not valid, but the act of articulating it allowed me to check it, find it wanting, and modify it. Notice the use of specializing:

- randomly, to get the feel of the question;
- systematically, to prepare the ground for generalizing;
- artfully, to test the generalization.

The resolution also displays several different uses of generalizing. Generalizing special results (the WHAT) led to the conjecture that any arrangement can be coloured by two colours. Generalizing methods led to the (false) 'opposite' rule, the (valid) 'adjacent' rule, and finally to the colouring technique based on introducing the lines one at a time. The convincing stage, which was only briefly indicated, involved even more generalizing with the emphasis on WHY.

I suggested before *Patchwork* that you write down your thoughts, feelings and ideas. You may have decided that it was not necessary for one reason or another, and so not bothered. If you did not write things down, then you missed an opportunity to learn something about yourself and about the nature of thinking. I recommend that you take the time to work through most questions conscientiously. If you did try it, I suspect that you did not find it easy. It does at first seem awkward and unnecessary, but a little self-discipline at this stage will reap rewards later. To make it easier to write helpful notes, I am going to be more specific about useful things to write down. This added detail or structure can then become a framework to assist your mathematical thinking. The framework will give continuing assistance if you get it inside you and make it your own. Otherwise it can provide only vague, temporary help.

The framework consists of a number of key words. As you use these words they become endowed with associations with past thinking experience, and through these associations they can remind you of strategies that worked in the past. In this chapter four key words are suggested, and in Chapter 2 they will be considerably augmented. The whole framework of key words is called a RUBRIC, following the medieval custom of writing key words in red in the margins of important books. The activity of writing yourself notes I call RUBRIC writing.

The four key words that I suggest you begin using in your notes and in your thinking are

STUCK!, AHA!, CHECK and REFLECT.

STUCK!

Whenever you realize that you are stuck, write down STUCK! This will help you to proceed, by encouraging you to write down why you are stuck. For example:

I do not understand . . .
I do not know what to do about . . .
I cannot see how to . . .
I cannot see why . . .

AHA!

Whenever an idea comes to you or you think you see something, write it down. That way you will know later what the idea was. Very often people have a good idea, but lose it subsequently and cannot recall it. In any case, it feels good to write down AHA! Follow it with

Try . . .
Maybe . . .
But why . . .

CHECK

- Check any calculations or reasoning immediately.
- Check any insight on some examples (specializing).
- Check that your resolution does in fact resolve the original question.

REFLECT

When you have done all that you can or wish to, take time to reflect on what happened. Even if you do not feel that you got very far, it helps to write up what you have done so that you can return to it freshly and efficiently at some later date. It is also the case that the act of summarizing often releases the blockage. There are several things worth noting particularly:

● write down the key ideas;
● write down the key moments that stand out in your memory;
● consider positively what you can learn from this experience.

I strongly recommend getting into the habit of RUBRIC writing when working on any question. You may wish to change the key words to suit yourself, but what is important is to develop rich associations with the words you use so that they conjure up the more detailed advice being offered in this and subsequent chapters. You cannot possibly memorize all the helpful advice that could be given. Instead of relying on someone else to get you unstuck with some timely advice, you can draw on your own experience. A RUBRIC is the means for drawing on that experience, and Chapter 7 discusses how the links between detailed advice and RUBRIC words are forged.

The RUBRIC should not be followed slavishly or dogmatically. Rather, with a little practice, the RUBRIC phrases will arise naturally, identifying what is to be done and suggesting what might be done. Sometimes you can be almost afraid to write down an idea in case you lose hold of it as it is forming itself in your head, so do not rush into writing. It is true though that having these standard key words automatically available helps in pinning ideas down. Conversely, try to avoid writing down scraps of ideas haphazardly in random places on a page. Everyone finds it hard at first to write RUBRIC notes, but those who persevere find it a real benefit.

Enough talking about RUBRIC! Investigate the next question using RUBRIC writing to record your resolution. Remember to specialize, and try to generalize from the specific instances.

Chessboard Squares

It was once claimed that there are 204 squares on an ordinary chessboard. Can you justify this claim?

TRY IT NOW – USE RUBRIC WRITING

STUCK?

➤ Usually one would say that a chessboard has 64 squares.
➤ What other squares are being counted?

➤ If you feel you are getting mixed up, and the situation is getting too complicated, specialize! Work on smaller boards.

➤ You must have a systematic way of counting the squares, but there are lots of ways of doing it. Find at least two different ways before carrying out one of them.

A resolution

What can it mean? I am STUCK because there are only 64 squares on the chessboards that I have seen, 8 rows and 8 columns. AHA! I get it, they are counting bigger squares as well, like these:

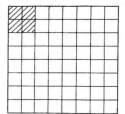 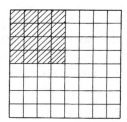 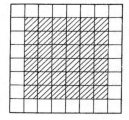

With this new interpretation of 'square', I WANT to count the number of 1 × 1 squares (there are 64 of these), 2 × 2 squares, 3 × 3 squares and so on up to the 8 × 8 square (there is only one). I've got to complete a table like this:

Size	1 × 1	2 × 2	3 × 3	4 × 4	5 × 5	6 × 6	7 × 7	8 × 8
Number	64							1

and show that the total is 204. At least it seems plausible!

TRY to count the 2 × 2 squares. Look at them; notice that they overlap.

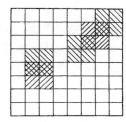

I will need a systematic approach to count these. How many touch the top line of the chessboard?

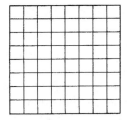

I count 7. How many touch the next line down? Count in the same way: 7 again. How many touch the next line? OOPS! What do I mean by a square 'touching' a line? I must mean that the top of the little square is on the line, otherwise I would count some squares twice. Counting gives 7 again. AHA! – there will be 7 on each row (generalizing). Along how many rows must I count? There are 9 horizontal lines altogether but the bottom two won't be touched by any 2×2 squares; 7 squares touching each of 7 rows gives 49 altogether.

Size	1×1	2×2	3×3	4×4	5×5	6×6	7×7	8×8
Number	64	49						1

AHA! I think I see a pattern here because 64 is 8×8 and 49 is 7×7. I guess then that there will be $36 = 6 \times 6$ of the 3×3 squares (generalizing and conjecturing).

CHECK by counting the 3×3 squares. How many touch the top line? I count 6 but now I see the reason (generalizing). There are 9 vertical lines intersecting the top line and each point of intersection can be the top left-hand corner of a 3×3 square – except for the three points on the extreme right. Therefore there are $(9 - 3)$ squares of size 3×3 touching the top line and, generalizing, there will be $(9 - K)$ squares of size $K \times K$ touching the top line. Also, there will be $(9 - 3)$ horizontal lines that the tops of the 3×3 squares can touch (9 altogether and the bottom 3 can't be used). Therefore $(9 - K)$ lines can be touched by $K \times K$ squares. Altogether there must then be $36 = (9 - 3) \times (9 - 3)$ squares of size 3×3, and there will be $(9 - K) \times (9 - K)$ squares of size $K \times K$.

Now I can fill in the table. Fortunately, I see that my previous results (for 1×1, 2×2 and 8×8) fit nicely and my guess is confirmed (CHECKING).

Size	1×1	2×2	3×3	4×4	5×5	6×6	7×7	8×8	$K \times K$
Number	64	49	36	25	16	9	4	1	$(9 - K)^2$

Pleased as I am with this table I am not finished yet! I WANT the total number of squares on the board which is

$$64 + 49 + 36 + 25 + 16 + 9 + 4 + 1 = 204$$

On REFLECTING, I notice that this result generalizes to chessboards with N rows and N columns. On such a board the number of $K \times K$ squares is found by noticing that there are $(N + 1 - K)$ squares of size $K \times K$ in any one row, and there are $(N + 1 - K)$ rows, so the number of squares of size $K \times K$ is $(N + 1 - K) \times (N + 1 - K)$. The total number of all sizes is then

$$(1 \times 1) + (2 \times 2) + (3 \times 3) + (4 \times 4) + \cdots + (N \times N)$$

The key idea was the systematic approach to counting the 2×2 squares. This idea generalized and produced the desired result. The transition from confusion and uncertainty about what to do with all those overlapping 2×2 squares, to the simplicity and calm of counting those that touched a line, stands out in my memory.

Now compare your notes with mine. There are many different ways of tackling this question, and you may have found that my system for counting the squares is quite different from yours. Counting them by marking the position of the centre of the squares, with a different colour for each size is a good way that leads to a nice geometric pattern reflecting the arithmetic result.

Look back carefully through your resolution to see where specializing and generalizing occurred. Notice particularly the different ways in which generalizing entered mine. The simplest occurrence was when I noticed that there would be the same number of 2×2 squares on each row. At a higher (or is it deeper?) level, while doing the 3×3 case I discovered the pattern connecting square size $K \times K$ with the number of squares in each row $(9 - K)$ and the number of rows (also $9 - K$). This enabled me to avoid doing the individual calculations for the 4×4, 5×5, 6×6 and 7×7 squares. Finally I was able to generalize my overall result when I saw how the size of the chessboard (8×8) entered the calculations, leading to the $N \times N$ board. Can you identify similar uses of generalization in your notes? My resolution is fairly typical of the results obtained by using the RUBRIC, although again it is rather more formal and complete than the notes made in the heat of the moment. Perhaps the main feature of my resolution is the large number of subsidiary questions which I found it useful to write down.

The idea of the RUBRIC, as I hope my resolution shows, is not that it be followed slavishly, nor that it interfere with your useful thinking. Rather it is a framework for organizing, recording and creating mathematical experiences. If it slows you down when your mind is racing, that may well be to the good. If it suggests things to do when you are stuck then that too is a benefit.

Review and preview

This chapter has introduced two fundamental mathematical processes: specializing and generalizing. There is no need to stare at a blank page, and no need to rush blindly at the first idea that arises. Anyone, when faced with a question, can try some specific example which brings the question into an area of confidence. It is no good trying examples which are themselves abstract and remote. The idea is to interpret the question through examples which are concrete and confidence-inspiring without trying to resolve the question itself. Then and only then can further specializing help to reveal a sense of what is going on. A resolution may then follow.

Trying to articulate a sense of some underlying pattern is called generalizing. It means noticing certain features common to several particular examples and ignoring other features. Once articulated, the generalization turns into a conjecture which must then be investigated to see if it is accurate. This whole process is the essence of mathematical thinking.

SPECIALIZING means choosing examples

- randomly, to get a feel for the question;
- systematically, to prepare the ground for generalizing;
- artfully, to test a generalization.

In the event that no pattern emerges, specializing means simplifying the question, making it more specific or more special until some progress is possible.

GENERALIZING means detecting a pattern leading to

● WHAT seems likely to be true (a conjecture);
● WHY it is likely to be true (a justification);
● WHERE it is likely to be true, that is, a more general setting of the question (another question!).

A format for writing notes about your thinking called a RUBRIC was suggested, to help you notice, record and thereby to learn from your experience of thinking mathematically. If it serves only to make your doodlings more coherent it will have made a significant impact. Its further potential will unfold as the book progresses.

The RUBRIC words introduced so far are

STUCK!, AHA!, CHECK and REFLECT.

The RUBRIC can be thought of as a scaffold around which a resolution is built. It also encourages checking and reflecting on your resolution, an essential ingredient for improving your mathematical thinking.

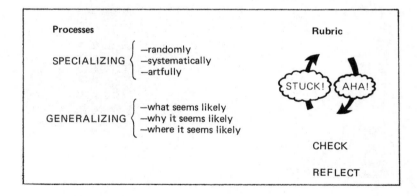

Many of the themes of later chapters have been hinted at in the discussion of the questions in this chapter. Chapter 2 aims to increase awareness of phases that take place in thinking, and augments the RUBRIC. It emphasizes the importance of spending time at the beginning getting to grips with the question, and at the end reviewing what has been done. The central phase of mounting an attack on a question is considered in subsequent chapters. All of them come back to, or are based on, the fundamental processes of SPECIALIZING and GENERALIZING.

You might like to consolidate these ideas by trying some of the questions in Chapter 10. The following are related to the questions in this chapter:

Ins and Outs	*Fred and Frank*
Divisibility	*Speed Trap*
Finger Multiplication	*Sums of Squares*

See Chapter 11 for other curriculum-related questions.

Reference

Banwell, C., Saunders, K. and Tahta, D. (1986) *Starting Points for Teaching Mathematics in Middle and Primary Schools*, updated edn. London: Oxford University Press.

2
Phases of work

In this chapter the process of tackling a question is divided loosely into three phases, called Entry, Attack and Review. Moving from one phase to another corresponds to a change in your feelings about the question and reflects the progress which is or is not being made. Learning to identify these phases in your own thinking will enable you to recognize appropriate activities.

Entry

Attack

Review

It may seem that, of the three phases, Attack should be most crucial since it encompasses the bulk of the obviously mathematical activity. However, quite the opposite is the case. Most people fail to resolve questions satisfactorily because of inadequate attention to Entry and Review. The Attack phase can only come about if the question has been satisfactorily entered, and if time has been devoted in the past to learning from experience by reviewing key moments in thinking. This chapter concentrates on Entry and Review, leaving Attack to later chapters.

Three phases

Think back to one of the questions presented in the last chapter, say *Warehouse*. Even after reading the question two or three times you may easily have felt that you had not really come to grips with the question and were still unsure of what was involved. Most people find that it takes some time and effort before they are sufficiently at ease with a question to begin an Attack. The initial Entry phase of tackling a question begins when I first encounter the question, and ends when I have become involved in attempting to resolve it.

Some people are so eager to begin that they jump at the first idea which comes along and rush into a full-scale assault without first taking time to survey the scene and assess what is involved. If the question fails to succumb to the initial onslaught (often because it has not been understood), then it is necessary to start afresh. Consequently it is worthwhile learning to begin effectively.

The major effort to resolve a question occurs in the Attack phase. This may lead ultimately to a complete resolution, or it may terminate in an incomplete resolution consisting of conjectures and unresolved questions. In either case, activity should not cease until after a final phase of Review, during which the work is checked, the processes and difficulties reflected upon, and the question and resolution extended where possible.

For example, with *Warehouse* my Entry phase consisted of activity leading to the conjecture that the order of calculation did not matter. In the Attack phase I tried to show that this conjecture was true for all prices, and in the Review phase I reflected on the way I had used specializing and extended the question to deal with various discount and interest rates.

The three phases grow quite naturally out of the fundamental processes of Chapter 1. Work in the Entry phase often begins with specializing in order to get to grips with the question. Grappling with the question is the Attack phase

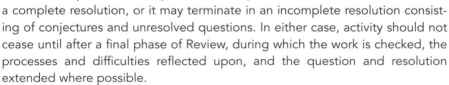

and involves more specializing and generalizing. This is where the most intense STUCK!s and AHA!s occur.

Attempts to resolve difficulties may stay within Attack or may lead back to Entry. Before leaving a question it is essential to carry out a third phase, Review. Discovery of an error or inadequacy may lead back to Entry or to Attack, and if an interesting new question is uncovered, perhaps through generalizing the resolution, the whole process begins again.

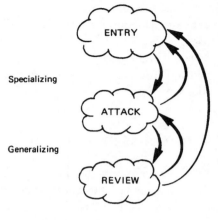

These three phases provide the backbone for further discussion of tackling questions. In this chapter I shall elaborate Entry and Review, suggesting additional words to add to your RUBRIC habit of writing down processes as well as ideas. The more complex phase of Attack is the subject of later chapters.

The Entry phase

It is important to recognize that an Entry phase can and should exist. Many people read a question once or twice and then expect to jump straight into the final solution, yet rarely is this possible. Work in the Entry phase prepares the ground for an effective attack and it is therefore essential that adequate time is devoted to it.

Entry begins when I am faced with a question. Usually the question has been written down, so most of the Entry phase advice can be summarized as **really read it**! In other cases the question presents itself, perhaps from work on another question or from a situation outside mathematics. Then the Entry phase work is largely in formulating the question precisely and in deciding exactly what I want to do. In either case the directions to take are fairly straightforward. I have to get to grips with the question in two ways: by absorbing the information given and by finding out what the question is really asking. The other activity which often takes place during Entry is to make some technical preparations for the main attack, such as deciding on a notation or a means of recording the results of specializing.

It is helpful, therefore, to structure work in the Entry phase by responding to three questions:

1 What do I KNOW?
2 What do I WANT?
3 What can I INTRODUCE?

You should begin to incorporate these questions into your RUBRIC writing. The order in which they are answered is not important, for they are closely linked. Bearing in mind KNOW, WANT and INTRODUCE particularly, try the next question.

Tethered Goat

A goat is tethered by a 6 metre rope to the outside corner of a shed measuring 4 metres by 5 metres in a grassy field. What area of grass can the goat graze?

TRY IT NOW

STUCK?
➤ INTRODUCE a diagram.
➤ Write down what you KNOW.
➤ Be clear on what you WANT.
➤ Break up what you WANT into manageable pieces.

My first instinct was to introduce a diagram and to mark the given information and its simplest consequences on it. A diagram is one of the most powerful tools for assimilating information, and it can be useful not only in geometric questions, but in a surprising variety of other situations. Having sorted out what was known, I found that relating KNOW and WANT suggested breaking up the total area into simpler pieces (a form of specializing) which could easily be worked out and added up.

The next sections give general advice for Entry in more detail.

Entry 1: What do I KNOW?

The *Tethered Goat* shows clearly that there are actually two aspects to what I KNOW. There is what I KNOW from the question, and what I KNOW from past experience. While reading I found information about the shed and the goat, and as I drew a diagram and marked information on it, facts and skills from past experience arose spontaneously inside me. In the absence of such spontaneous resonance, I find it useful to ask myself if I have ever seen anything similar or analogous and this often throws up an idea. The most effective way to stimulate resonances from the past is to have employed the RUBRIC approach of making process notes, and to have reflected upon them in the final phase.

In the Entry phase, both aspects of 'What do I KNOW?' are answered by extracting from the problem statement all the relevant information that it supplies, thoroughly digesting it and writing down ideas that seem relevant. Since mere acquaintance with the information is hardly ever sufficient to resolve a question I will suggest several simple means of assimilating the information provided.

'Read the question carefully' is obvious advice, but it is frequently neglected. Not only is information ignored, but the rush to get going on the question often means that the real question is misunderstood or missed altogether. The impact of many jokes, riddles and puzzles relies wholly upon a tendency to misread, as illustrated by the following classic:

> As I was going to St Ives
> I met a man with seven wives.
> Each wife had seven sacks,
> Each sack had seven cats,
> Each cat had seven kits.
> Kits, cats, sacks and wives
> How many were going to St Ives?

Calculations are unnecessary if you read carefully to find out the direction of travel of the parties. Another favourite is:

> How much dirt is there in a hole 3 ft 6 inches wide, 4 ft 8 inches long and 6 ft 3 inches deep? Metric enthusiasts can take it as 1.06 m by 1.42 m by 2.01 m.

Many people start multiplying the numbers together!

A second useful way of absorbing the information provided by a question is specializing, introduced in Chapter 1. In both *Paper Strip* and *Palindromes* specializing was used to develop familiarity with the question, namely to discover the exact form of a four-digit palindrome or to understand the rules for folding the paper strip. The purpose of specializing is to get a sense of what the question is about, and to gain confidence and familiarity with the objects it involves.

One of the best tests that you can apply to see if you have absorbed what a question tells you is to write down or tell someone else the question in your own words. This does not mean that you should memorize the question or copy it down, but rather that you should seek out the essence and write it down in your own way. If you are completely familiar with the information then you should be able to reconstruct the question.

TRY RESTATING THE ESSENCE OF TETHERED GOAT

I restated *Tethered Goat* this way: the goat swings around the shed in a sequence of circular arcs. I want to find the total area of the sectors. The dimensions themselves are not really critical.

Sometimes the information provided in the statement of a problem is long and complicated and no one could be expected to reconstruct it in detail. In such cases, it is worthwhile spending some time classifying, sorting and organizing the information.

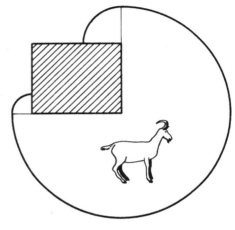

A good method of doing this is to construct diagrams or tables which display it concisely and systematically. The aim is to build up a sense of the type of information that has been provided so that when the problem is being attacked, appropriate pieces of information can be readily selected and used.

Look at the information provided in the next question and try to put it into a convenient form, getting a sense of what each sentence contributes before actually trying to resolve the question.

Ladies Luncheon

Five women have lunch together seated around a circular table. Ms Osborne is sitting between Ms Lewis and Ms Martin. Ellen is sitting between Cathy and Ms Norris. Ms Lewis is between Ellen and Alice. Cathy and Doris are sisters. Betty is seated with Ms Parkes on her left and Ms Martin on her right. Match the first names with the surnames.

DO IT NOW

STUCK?

➤ INTRODUCE a diagram of the circular table.
➤ Write down on a diagram or in a chart what you KNOW about the neighbours of each woman.
➤ What information have you not used?

A resolution

The information on the neighbours of Ms Osborne, Ellen, Ms Lewis and Betty enable the seating positions to be found, although with mixed first names and surnames. The information that Ms Parkes is on Betty's left serves to give an orientation to the seating arrangement. When checking what the five names and five surnames are, it becomes apparent that the sentence 'Cathy and Doris are sisters' contributes only Doris's name.

This is what is meant by getting a sense of what you know: examining each statement carefully to find its likely impact on the resolution.

Questions like *Ladies Luncheon* are very common in collections of puzzles, and organizing the information on diagrams and charts is often a useful technique for solving them. Puzzles of this kind are valuable because they provide experience of being systematic in the use and recording of information. This is often important in the Attack phase of more significant or difficult questions. Having accumulated a lot of information and having done several examples, it helps to pause and organize it, writing it down coherently.

Extracting and absorbing information is important thinking behaviour at any time. Recently I bought a new sewing machine. Browsing through the instruction book was useful but I found that I began to understand the instructions fully only when I adopted a more active approach and sat down at the machine with the book, material and thread. Exactly the same situation prevails in mathematics. It is the active approach of drawing diagrams, specializing, constructing tables, reformulating the question and trying things out that leads to success.

Entry 2: What do I WANT?

'What do I WANT?' directs my attention to the task of finding out what it is I have to do

> to find an answer

or

> to prove that something is true.

Failure to clarify exactly what has to be found or proved is a major cause of difficulties in resolving questions, yet the victim is usually unaware that this is the root of the trouble. Conscientious use of RUBRIC will reduce this. For example, in a question like *Chessboard Squares* which provokes numerous subsidiary questions such as

> How many 2 × 2 squares are there?

or

> How many 2 × 2 squares have their tops on this line?

they can easily slip away if they are not recorded as what I **currently** WANT.

Sometimes it is easy to answer 'What do I WANT?' In such cases, for example with *Tethered Goat* or *Paper Strip*, the two activities of clarifying what has to be done and absorbing the information are closely linked. When, as with *Tethered Goat*, it is a number that has to be found, introducing a symbol for it such as A or AREA helps simply by giving a concise name to what is otherwise described in a sentence. It is important to specify precisely what the symbol stands for, in order to avoid possible confusion later, in this case between the total area and the area of individual sectors. So too with *Paper Strip*, being clear about what you are counting is essential!

In a question like *Patchwork*, understanding exactly what is required is rather subtle even though (or perhaps because) the question states it quite concisely. Specializing can help find out what the requirements are.

Careful reading is again important. With 'What do I WANT?' it is particularly important to be alert to possible ambiguities or misinterpretation of the question. Consider this example from *Amusements in Mathematics* by Dudeney (1958) which hinges upon the interpretation of 'fractional part' and 'exceeds'.

Fractious

'Look here, George,' said his cousin Reginald. 'By what fractional part does four-fourths exceed three-fourths?'

'By one-fourth!' shouted everybody at once.

'Try another one,' George suggested.

'With pleasure, when you have answered that one correctly,' was Reginald's reply.

TRY IT NOW

Reginald wanted the answer one-third, because three of anything (in this case, quarters), if increased by one-third, become four. The others were answering the questions as if the phrase

of a cake

had been inserted in the question

after 'fractional part',
after 'four-fourths',

and

after 'three-fourths'.

This example can perhaps be dismissed as just a badly posed question but cases of genuine ambiguity, misinterpretation and lack of sufficiently clear definition are frequent in mathematical thinking. For example, asked for the number of squares on a chessboard, 64 and 204 are both good answers, depending on different interpretations. It is valuable to learn to notice possible ambiguity and multiple interpretations. Often, deciding upon a precise definition of a concept that was previously considered obvious has been the fundamental step in opening up a new area of mathematics. For example, the intuitive idea of a continuous curve, when subjected to close scrutiny spawned an entire branch of mathematics.

Answering 'What do I WANT?' is not always as easy as it sounds. When the question has arisen naturally from the resolution of another question, deciding exactly what you want to do can be a major task, requiring careful thought. Similarly, a question from life like the next one can be resolved in many different ways, depending largely on what it is seen to be asking.

Envelopes

I have just run out of envelopes. How should I make myself one?

TRY IT NOW

STUCK?

➤ Have you examined some typical envelopes?
➤ What size paper do I need to make an envelope for a given size of writing paper?
➤ Do I need an envelope at all?
➤ What properties must an envelope have?

There are a great many avenues to explore, depending on what question strikes you. I have often wondered at the range of sizes, shapes and styles of envelopes. Finally I took one apart and was delighted by the shape. I then wondered whether it was the most efficient shape and so on. On the other hand, there is no need to be fancy if I suddenly find myself without an envelope. A fold, a bit of tape, and the question is resolved!

Entry 3: What can I INTRODUCE?

Both *Tethered Goat* and *Ladies Luncheon* cry out for diagrams. Very often it is necessary to introduce other elements such as charts or tables for organizing data and symbols to stand for various objects. Frequently there are objects not explicity referred to which may need names as well. Sometimes a question which seems awkward or obtuse becomes transparent when you transfer it to a new context. Diagrams for problems like *Tethered Coat* and *Ladies Luncheon* help by extending your mental screen, and by forcing you to extract the essential features involved. In *Paper Strip* (Chapter 1), manipulating paper is much easier than doing it in your head, though it is good exercise to try it mentally first before resorting to another representation. The mental work prepares the ground for a diagram and, conversely, manipulating physical objects with attention assists your ability to manipulate mental images.

It is easy to know all this in theory, but it is not always so easy in practice. It often seems difficult to go beyond the confines of the question, but additional objects or ideas do turn up when what you KNOW and what you WANT are translated into a context appropriate to your experience and confidence. Adding INTRODUCE to your RUBRIC vocabulary is intended to foster a mental attitude of freedom to introduce helpful elements. You are in charge of the question, not the other way round! To be more precise about what sorts of things can be of assistance, it is useful to distinguish between them:

Notation:	choosing what to give a name to, and what name to give.
Organization:	recording and arranging what you KNOW.
Representation:	choosing elements that are easier to manipulate and substituting these for the elements in the question.

The next question illustrates all three aspects of INTRODUCE

Cubes Cubed

I have eight cubes. Two of them are painted red, two white, two blue and two yellow, but otherwise they are indistinguishable. I wish to assemble them into one large cube with each colour appearing on each face. In how many different ways can I assemble the cube?

TRY IT NOW – FIRST WITH IMAGINARY CUBES,
THEN WITH ACTUAL CUBES

STUCK?
- ➤ Specialize – can you find one arrangement?
- ➤ What do you KNOW? Are the conditions clear?
- ➤ What do you WANT? When are two large cubes different?
- ➤ What presentation can you INTRODUCE? Before using eight painted cubes try drawing a cube or diagrams of its faces, or try looking at a convenient box – it may help you visualize the situation even if you do not touch it.
- ➤ INTRODUCE a way of describing/recording different solutions.

Reflecting, I find that questions involving three dimensions are almost invariably easier to resolve using physical objects. The main steps here seem to be finding one solution, noticing the pattern of colours and then deciding upon the meaning of 'different'. I decided that two cubes are different if there is no way that one can be turned around so that the pattern of colours on it matches the pattern on the other. Different definitions of 'different' may of course result in different answers. If actual cubes are used as a representation to think about this question, it is particularly important that the various solutions obtained are recorded carefully (physically, pictorially or symbolically) so that they can be compared. Later when I checked the resolution a convenient notation for solutions was essential. Thus all three aspects of INTRODUCE (representation, notation and organization) are important in resolving this question.

It is unfortunate that many adults seem to feel that using something concrete to help resolve a question like *Cubes Cubed* or *Paper Strip* (Chapter 1) is childish and not an acceptable technique for an adult, who ought to be able to 'think abstractly'. In fact, the use of an appropriate representation, even a very simple one, can often turn an apparently difficult question into an easy one. Remember that the aim of a mathematical thinker is to get a good resolution – not to do the question the hard way. Anything that helps can and should be used. Just looking at a box in the corner of the room can be of considerable assistance for *Cubes Cubed*, even if it is not touched. Somehow, looking at the box extends the mental screen, increasing powers of visualizing and mentally manipulating objects. A similarly powerful use of a simple representation is often observed with the next question, *Quick and Toasty*, which is usually first tackled abstractly. Using pieces of paper to represent slices of bread frequently helps people solve this question quickly. Again the presence of the paper, even if it is not touched, seems to extend powers of visualizing and makes it easy to consider a range of possibilities.

Quick and Toasty

Three slices of bread are to be toasted under a grill. The grill can hold two slices at once but only one side is toasted at a time. It takes 30 seconds to toast one side of a piece of bread, 5 seconds to put a piece in or take a piece out and 3 seconds to turn a piece over. What is the shortest time in which the three slices can be toasted?

TRY IT NOW

STUCK?

➤ Watch out for ambiguity!
➤ It can be done in less than 140 seconds.
➤ Which time periods should you concentrate on minimizing?

When you have found an efficient method of toasting three pieces of bread, try to extend the result to toast more slices using grills with larger capacity. The general result exhibits several interesting patterns.

Entry summarized

It has been suggested that work in the entry phase is based on answering three questions:

1 What do I KNOW?
2 What do I WANT?
3 What can I INTRODUCE?

Careful reading (to avoid overlooking information and to notice ambiguities) and specializing are obvious suggestions for answering both 'What do I WANT?' and 'What do I KNOW?' It is important to obtain at least a strong sense of the type of information that the question contains and how it might be used. Trying to reconstruct the question (not necessarily in detail) is a useful test to see if you have a real sense of what you WANT. Writing down the essentials in your own words can be particularly helpful whereas copying out the question as stated is usually a waste of time. INTRODUCING diagrams, symbols and charts can substantially help to get into the question and INTRODUCING a notation, a means of recording or a representation, puts you in a good position to begin the Attack phase.

This might be a good time to try *Jobs*, *Cycling Digits* and *Glaeser's Dominoes* in Chapter 10, paying particular attention to INTRODUCE.

The Attack phase

Thinking enters the Attack phase when you feel that the question has moved inside you and become your own. The phase is completed when the problem is abandoned or resolved. The mathematical activities which might take place in Attack are complex and varied and will be described in detail in the following four chapters. The states which are particularly associated with Attack are STUCK! and AHA! and the fundamental mathematical processes called upon are conjecturing (Chapter 4) and justifying convincingly (Chapter 5). These in turn depend on specializing and generalizing.

During Attack several different approaches may be taken and several plans may be formulated and tried out. When a new plan is being implemented, work may progress at a great rate. On the other hand, when all ideas have been tried, long periods of waiting for new insight or for a new approach may characterize the phase. These periods of waiting and mulling are the topic of Chapter 6.

For the time being, concentrate on recognizing when you are stuck, and accepting it calmly, without judgement or tension. After all, it is only from being stuck and accepting it, that it is possible to learn how to get unstuck. Much of that learning comes about as a result of the Review phase, which is all too frequently overlooked.

The Review phase

When you reach a reasonably satisfactory resolution or when you are about to give up, it is essential to review your work. As the name suggests it is a time for looking back at what has happened in order to improve and extend your thinking skills, and for trying to set your resolution in a more general context. It involves both looking back, to CHECK what you have done and to REFLECT on key events, and looking forward to EXTEND the processes and the results to a wider context. It is worthwhile adding EXTEND to your RUBRIC writing, to complement CHECK and REFLECT, and the three words together help to structure the Review phase activities:

1 CHECK the resolution
2 REFLECT on the key ideas and key moments
3 EXTEND to a wider context

The best way to get benefit from Review is to write up your resolution for someone else to read. Bearing the three activities in mind, go back to your notes on *Chessboard Squares* and write them up coherently so that someone who has not thought about the question can follow **what** you have done and **why**. In so doing you will probably come up with some ideas for improving your resolution and extending it to solve other problems.

DO IT NOW

If you did go back over your notes conscientiously, you will recognize the contrast between your appreciation of *Chessboard Squares* and of *Palindromes* which probably did not receive such attention. By converting the relative chaos of your notes, complete with their RUBRIC comments, into an understandable exposition of what you did and why, you not only end up with a sense of satisfaction at having a finished product, but you are also brought into close contact with the key events. The central ideas of your resolution have to be identified and distinguished from those of minor importance, so REFLECTING has already begun. Keep it going! Writing down anew the details of the resolution and reconstructing the arguments ensures a thorough CHECK. By going over the details and trying to clarify them for someone else you prepare the ground for EXTENDING your understanding of the question and its implications. Try to bring yourself to do this writing up on the remaining examples in this chapter and in this book. Self-discipline is required, but necessary if you really want to improve your thinking, and it is amply rewarded.

Review 1: CHECK the resolution

In Reviewing *Chessboard Squares* I found that I had several kinds of CHECKING to carry out. I had to:

- CHECK the arithmetic and the algebra for mistakes in computation;
- CHECK the arguments to make sure that the computations did what I thought they were doing;
- CHECK the consequences of conjectures to see if they were reasonable ($9 - K$ squares of size $K \times K$ touching the top line means that for $K = 8$ there should be one square size of 8×8, which checks);
- CHECK that I had actually answered the original question and not just a subsidiary one.

The first two kinds of CHECKING were done at the time, but CHECKING in the heat of the moment is less reliable than later when it can be more relaxed and dispassionate. In any case, looking for errors by repeating exactly what has been done is a poor way to CHECK, as anyone knows who has tried to balance a cheque book. It is much better to proceed in a different way. With the hindsight of having reached a resolution, new and simpler ways may occur to you, thus clarifying your resolution and adding to your insight. For example, instead of locating the top edge of a square in *Chessboard Squares*, focusing on the centres of the squares leads to geometrical patterns with arithmetic content.

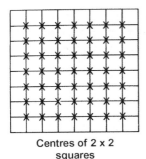

Centres of 2 x 2
squares

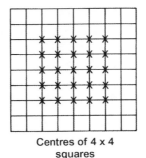

Centres of 4 x 4
squares

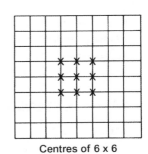

Centres of 6 x 6
squares

The second type of checking, CHECKING consequences, is a powerful device. You may recall from *Palindromes* that it was by CHECKING consequences of my conclusion that I discovered my error. Successive palindromes always differing by 110 would mean that all palindromes have to end in 1 which is ridiculous!

Taking the advice to write up your resolution tidily, to seek a new path and to look for consequences of results will help in checking a resolution, but how can you be absolutely certain that no errors have been made? In the last analysis, you cannot be sure. There have even been results and proofs that the mathematical community has accepted as correct for many years, yet mistakes in them have subsequently been found. CHECKING therefore is a difficult task and so it is taken up again in Chapter 5 on Convincing.

Review 2: REFLECT on the key ideas and key moments

REFLECTING is possibly the most important activity for improving mathematical thinking. Contrary to the cliche, I do not learn from my experience; that is, not unless I reflect on what I have done. So that REFLECTING does not become day-dreaming I suggest you structure it by identifying key ideas and key moments in the resolution. For example, in *Chessboard Squares* the key ideas were realizing that I could count the 2×2 squares by concentrating on the number of them touching a given line, and noticing the pattern of square numbers which arose (64, 49, . . .). Key moments were

- realizing the need for systematic counting;
- using further specializing to check the squares pattern conjecture (64, 49, . . .);
- specializing to see why the generalization was correct.

These stand out sharply in my memory. Next time I find myself counting patterns, the *Chessboard Squares* system will come to mind because I took the time to recall it during my Review. Next time I want to see why a generalization might be correct, I will try further specializing.

Recalling the **key ideas** and investing them with vividness is the way to build up your experience, your bag of mathematical tricks. Having a written account of your thinking using RUBRIC or similar words is very helpful in recalling **key moments**, and you can learn to 'photograph' them so that they will be preserved with their original clarity. The photographs can then become your personal tutor, helping you when you are stuck by recalling memories of key ideas that have worked in the past. Taking these snapshots and using them to recall relevant experience is the subject of Chapter 7.

Review 3: EXTEND to a wider context

Think back to *Chessboard Squares* again. As I was reviewing my work, I saw the relationship between the width of the chessboard (say *N* squares) and the number of squares of size $K \times K$ that fit along a row, namely $N + 1 - K$. As I began to understand why this is true, I realized that this result can be used to predict the number of squares on a 'chessboard' of any size. Such an EXTENSION is not forced, but arises naturally from my deepening understanding. EXTENDING goes hand in hand with REFLECTING. For example, trying to extend the result of *Palindromes* to palindromes with other than four digits highlights the dependence of the resolution on four being an even number. The extension is provoked by a question of the form:

Why four? What if . . . ?

As I look back to see what was important, understanding grows and my previous struggle will sometimes yield an unexpected richness of results with little extra effort. The following question, for example, is easily cracked after *Chessboard Squares*.

Chessboard Rectangles
How many rectangles are there on a chessboard?

TRY IT NOW

STUCK?
- ➤ What do you WANT?
- ➤ Try a small chessboard first (specialize).
- ➤ What systematic way of counting the rectangles will be best?
- ➤ Examine the method used to count the squares on a chessboard and generalize.
- ➤ EXTEND? Generalize the chessboard!

Now the isolated answer 204 for the number of squares on a chessboard has been set into a wider context. It is a special case of a broader pattern. One of the qualities of an interesting question is that it has several EXTENSIONS which broaden the scope of the original. You only really understand a result when it fits into a broader context. Very often this comes about by removing or relaxing assumptions, for example:

Why just an ordinary chessboard? Try $N \times N$ squares.
Why just count squares? Count rectangles.
Why start with a square? Count rectangles in a rectangle.

and then

Why only count squares lined up with edges parallel to the original?
Why work only in two dimensions?

and so it goes on. *Triangular Count* in Chapter 10 points in yet another possible direction for extending *Chessboard Squares*.

Paper Strip (page 3) from Chapter 1 is worthy of EXTENDING, if only to remove that sense of 'I found the answer but so what?' In the light of the Entry and Review advice, this would be a very good time to do more work on *Paper Strip*. What happens if it is folded in thirds each time? What happens when you unfold the creased strip and examine the pattern of creases?

Putting a question in a more general context, that is generalizing, reveals its significance in a wider scheme of things. There is another advantage of EXTENDING a resolution. Mathematical thinking does not begin until you are engaged by a question. The most engaging question is always your own, either because you made it so by specializing and Entry activity, or because it arose from your experience. Indeed, it is curious that some of the most interesting and challenging questions have arisen from attempting to generalize apparently boring looking results. EXTENDING is a good source of your own questions. Chapter 8 deals with the more general task of becoming your own questioner. For one way to extend *Paper Strip* see *Ins and Outs* in Chapter 10.

Practising Review

Have a go now at the next question, and do a complete RUBRIC write-up for it. The question itself is easily answered, so concentrate on the Review.

Creepy Crawlies

Ross collects lizards, beetles and worms. He has more worms than lizards and beetles together. Altogether in the collection there are 12 heads and 26 legs. How many lizards does Ross have?

STUCK?

➤ Specialize.
➤ What do you KNOW?
➤ What do you WANT?
➤ What can you INTRODUCE to help get started?
➤ How many unknowns? How many equations? What could help?

A resolution

I WANT to find the number of lizards. I KNOW there are 12 heads. AHA! That means there are 12 animals. I also KNOW there are 26 legs. Now lizards have 4 legs each (I shall suppose all the specimens are in good health) and beetles have 6 legs each and worms have none. So the 26 legs come from the beetles and lizards. For example, if there were only one lizard, there would be 22 legs on the beetles (specializing). AHA! This is impossible as it would mean

$$22/6 = 3 + 2/3 \text{ beetles.}$$

I could now try 2 lizards. This means there are 18 legs on the beetles, so there are 3 beetles and therefore 7 worms. This answers the question, provided I can show there are no other solutions. STUCK!, because I cannot think of a method. Should I try to INTRODUCE symbols and get some equations? AHA! I don't have to do that. With 7 lizards there would be too many legs so 6 is the most there can be. Now I can test each case in a table. There is only one possibility!

Number of lizards	1	2	3	4	5	6
Number of beetles	$3\frac{2}{3}$	3	$2\frac{1}{3}$	$1\frac{2}{3}$	1	$\frac{1}{3}$
Number of worms		7			6	
More worms than lizards and beetles?		YES			NO	

The Attack is over. I have checked arguments and calculations, and I have answered the original question. Now I will Review my resolution. For convenience, write the number of lizards as L, the number of beetles as B and the number of worms as W. Undoubtedly the key idea is that L, B and W all have to be positive whole numbers because all the creepy crawlies must be complete. Specializing brought this home to me. Being systematic about my specializing resolved the question, but would this always be a good method? Probably not; if the numbers involved were large (say there were 260 legs instead!) or the difference between the number of legs was known instead of their sum, then another method might have to be devised. Perhaps, though, there might then be a pattern amongst the values of L that give acceptable values for B. I do notice a pattern connecting L and B in my table. As L increases by 1, B decreases by 2/3 so in the B row every third column has a whole number. Perhaps such a pattern could be exploited to solve a similar problem involving larger numbers. I also notice that I was lucky that worms have no legs. This let me concentrate on L and B alone, then calculate W later. What if the worms were replaced by spiders?

REFLECTING upon my resolution has in this case raised the following new questions:

(i) How could I solve a similar problem with 26 and 12 replaced by much larger numbers?

(ii) How could I solve a similar problem with lizards, beetles and spiders?

Both these questions are worthy of investigation and they arose by critically reviewing the resolution of the original question. As yet they are not well posed because they have each arisen naturally from the resolution to another question.

Therefore much of the Entry phase for the new question will be directed towards formulating it more precisely. Answering 'What do I WANT?' is what you should do first.

TRY ONE NOW

There is one further important function of the Review phase. Being able to recognize analogous questions is an important weapon for thinking mathematically. One way of building this strength is to pause after resolving a question to consider how the same techniques might apply in other situations. Try to distinguish between the key ideas and the superficial aspects of the question. Clearly in the present case there are many physical contexts other than lizards and beetles which would lead to similar questions. For example, I could be told the total number of wheels on some cars and motorbikes in a parking lot, or the total face value of some stamps of known denominations. A more searching generalization would involve enquiring about what sorts of information, similar in kind to the lizards and beetles data, would nevertheless result in a unique solution.

Despite the rewards which the Review phase has to offer, it is often neglected. Why is this? After the initial flush of excitement at a successful Attack, there often arises a feeling of anticlimax when confronted with having to check a result that you already firmly believe. Also, if you have not really accepted the question as your own, there is a temptation to leave it as soon as possible and get on with something else, perhaps for more marks. This means that a valuable opportunity for learning about the processes of thinking is lost, together with a chance to learn more about mathematical content.

Review summarized

I have suggested that work during the Review phase is crucial to developing your mathematical thinking. It is based on three closely related activities:

1 CHECK the resolution;
2 REFLECT on the key ideas and moments in the resolution;
3 EXTEND the result to a wider context.

Begin the Review by writing up your resolution (however partial) for someone else to read. Doing this will automatically involve all four types of CHECKING, especially if you try to find a new path, and it will highlight the key ideas that were involved. Trying to recall and mentally 'photograph' these key moments is important for building up a useful reservoir of mathematical experience. EXTENDING a resolution will usually arise naturally, either as you become interested in pursuing a new aspect or as your understanding suggests further applications of what you have discovered. REFLECTING is central to the whole Review phase.

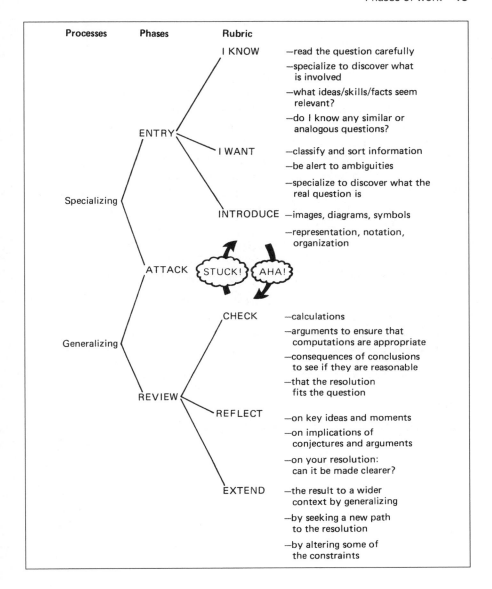

Processes	Phases	Rubric	
Specializing	ENTRY	I KNOW	—read the question carefully
			—specialize to discover what is involved
			—what ideas/skills/facts seem relevant?
			—do I know any similar or analogous questions?
		I WANT	—classify and sort information
			—be alert to ambiguities
			—specialize to discover what the real question is
		INTRODUCE	—images, diagrams, symbols
			—representation, notation, organization
	ATTACK	STUCK! AHA!	
Generalizing	REVIEW	CHECK	—calculations
			—arguments to ensure that computations are appropriate
			—consequences of conclusions to see if they are reasonable
			—that the resolution fits the question
		REFLECT	—on key ideas and moments
			—on implications of conjectures and arguments
			—on your resolution: can it be made clearer?
		EXTEND	—the result to a wider context by generalizing
			—by seeking a new path to the resolution
			—by altering some of the constraints

The three phases summarized

The phases of thinking about a question are not crisply distinct. They tend to be fuzzy around the edges because they are concerned with qualities of experience and not with mechanical activity. Work in one phase may easily lead back to a 'prior' phase or on to a final writing up. By learning to distinguish the salient qualities of each phase you can provide yourself with things to do which need doing when you find yourself momentarily stuck. If these suggestions become a habit for you, then a lot of aimless day-dreaming and unproductive thinking can be avoided.

The RUBRIC words, which were extended in this chapter as shown in the figure on the previous page, are generally indicative of the current phase. For example, if you find yourself in the depths of a question and suddenly find it necessary to go over what you KNOW and WANT, then you have in one sense re-entered the question, but of course with a lot more experience behind you than when you first began.

Behind all the activities described in the three phases lie the twin processes of specializing and generalizing. By specializing you find out what you KNOW, what you WANT, and what you might sensibly INTRODUCE. By specializing you uncover patterns leading to a generalization. By generalizing you come up with conjectures which can be CHECKED by further specializing and you can EXTEND the question to a wider context. The picture of thinking mathematically has now grown to include processes, phases and RUBRIC writing.

The following questions in Chapter 10 would be worth looking at now:

Jobs	*Cartesian Chase*
Glaeser's Dominoes	*Nullarbor Plain*
Cycling Digits	*Rolling Coins*
Diagonals of a Rectangle	*Odd Divisors*

See Chapter 11 for other curriculum-related questions.

Reference

Dudeney, H. (1958) *Amusements in Mathematics.* New York: Dover.

3
Responses to being STUCK

Everyone gets stuck. It cannot be avoided, and it should not be hidden. It is an honourable and positive state, from which much can be learned. The best preparation for being stuck in the future is to recognize and accept being stuck now, and to reflect on the key ideas and key moments which begin new and useful activity.

This chapter contains two questions on which to practise Entry activities, before launching into the details of Attack. I hope that you will get happily stuck, and learn from it!

Being STUCK

There are several different feelings connected with being stuck. For instance I can find myself

- staring at a blank page, at the question, or into space;
- resisting a computation or some other action;
- growing tense or even panicking because I cannot make progress;
- feeling frustrated because nothing seems to be working;

to mention only a few. In my experience, I have usually been stuck for some time before I become aware of it. At first the awareness is hazy and indistinct. It slowly grows until I am definitely both stuck and aware of being stuck. Only when I feel stuck and I am aware of my feelings can I take action. That is why RUBRIC writing is so useful, and particularly the writing of STUCK! or something similar. The act of expressing my feelings helps to distance me from my state of being stuck. It frees me from incapacitating emotions and reminds me of actions that I can take.

So what can be done about being stuck? Having recognized and accepted that you are STUCK, you can either stop working altogether, take a short break or keep going. Although it is often inviting to give up when you are stuck, it is not always the best idea. Good ideas often come just when it seems most hopeless. If you are going to take a break, remember first to record as clearly as possible what it is that you think is blocking you. There will be more on this topic in Chapter 6.

Now what do I do if I feel like continuing to work? Chapter 2 focused on the kinds of question which a tutor would use to help you, but they only become really useful when they are integrated into your personal experience of mathematical thinking. You then have your own internal tutor. In any case, the most sensible activity is to return to the Entry phase and to reassess:

1 What do I KNOW?
2 What do I WANT?
3 What can I INTRODUCE?

They suggest:

- summarizing everything that is KNOWN and WANTED;
- representing the question in some form that is concrete and confidence inspiring;
- taking advantage of the specializing already carried out;
- rereading or redigesting the question while looking for alternative interpretations.

Reading the question again carefully is not a sign of inadequate reading the first time. On the contrary, it is often the case that the question only really makes sense after you have specialized with various examples, so that you bring more relevant experience to the reading. Intentionally going back and rereading is the mark of someone gaining confidence in their thinking abilities and becoming aware of the process of thinking. Of course, freshly RE-ENTERING does not mean mindless rereading, over and over, but bringing the recent experience to bear on the reading, looking for alternative interpretations. Since you have not yet resolved the question there is still a gap between KNOW and WANT. By RE-ENTERING you are trying to find out just what you do KNOW, and how far it is from what you WANT.

Suppose now that you have conscientiously RE-ENTERED, that you have summarized everything you KNOW and WANT in your own words, and still there is a blank, a yawning chasm. Almost certainly what is wanted then is more extreme specializing. The purpose of specializing is to turn to confidence-inspiring examples from which a pattern can emerge. It may be that the question needs more drastic simplification. For example, in *Chessboard Squares* (Chapter 1) I could have decided to count the number of squares in a 2×2 'chessboard', then a 3×3 and so on. I could even have considered a 1×8 'chessboard', and more drastically still, 1×1, then 1×2 then 1×3 boards. By considerably simplifying the task of counting, I make space for pattern seeking, for seeing what is really going on.

I have concentrated on being stuck because most people only become aware that they need help when they finally realise that they are stuck. As soon as an idea comes, they are off and running, heedless of advice. It will help if you cultivate the habit of recording what you are trying to do or what you think might work, however hazy:

- to slow down a little when in full flight;
- to evaluate the ideas more fully and system-atically;
- to decode later what it is you thought you were doing.

It is for this reason that RUBRIC writing is recom-mended. Despite the resistance we all feel to slow-ing down an exhilarating flow of ideas, experience shows that the headlong rush to get to an answer is a primary cause of becoming truly STUCK! Learn to savour the resolution of a question in the way a gourmet savours his food, rather than gulping it down as quickly as possible! So, when you get an idea which feels good, try writing AHA!, and then recording the idea. If nothing else, writing AHA! will extend the sense of pleasure and satisfaction of having had an idea.

The next question is intended to take more than just a few minutes. It is well worth devoting sufficient time to it to experience both getting stuck and getting unstuck!

Threaded Pins

A number of pins are placed around a circle. A thread is tied to one pin, and then looped tightly around a second pin. The thread is then looped tightly round a third pin so that the clockwise gap between the first and second pin is the same as the clockwise gap between the second and third pin as illustrated in the example.

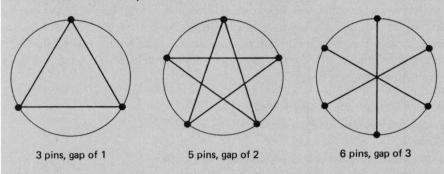

3 pins, gap of 1   5 pins, gap of 2   6 pins, gap of 3

The process is continued, always preserving the same clockwise gap until the first pin is reached. If some pin has not yet been used, the process starts again.

Five pins with a gap of two use just one thread, while six pins with a gap of three use three threads. How many pieces of thread will be needed in general?

TRY IT NOW

STUCK?

Entry

➤ The best advice is to specialize.

➤ Organize the results of specializing.

➤ What could you INTRODUCE to be able to express what you WANT succinctly?

Attack

➤ Try the subsidiary question: what pins can I reach from the starting pin?

➤ Make a conjecture, however wild.

➤ Now check your conjecture, looking for why it is right/wrong.

➤ You may find yourself making and modifying several conjectures before you find one succinct statement that covers all cases.

Review

➤ Even if you get totally stuck, go over what you've done **before** reading my resolution!

Other people's resolutions are tedious compared with your own. I strongly recommend **not** reading my resolution to *Threaded Pins* until you have solved it or become STUCK and have tried all the advice in Chapters 1 and 2.

A resolution

Having tried a number of examples on diagrams, I turned back to the question. I WANT some way of telling how many threads are needed when I KNOW the number of pins and the gap size. I must be systematic, but how can I deal with both the number and the gap changing simultaneously (organization)? AHA! I shall use a table. What do I WANT the entries to be? Ah yes, the number of threads for various combinations of pins and gap size. By doing examples on circle diagrams I get the results shown below.

Pins \ Gap	1	2	3	4	5	6	7	8	9
3	1	1	3						
4	1	2	1	4					
5	1	1	1	1	5				
6									
7									
8									

I am fully involved now in completing the table. I am so involved that I have forgotten what the question is. Rereading it, I see that I WANT to be able to predict how many threads are needed for a given number of pins and for a given gap size. I must INTRODUCE names for the number of pins and the gap size. For the moment I will use *pins* and *gap* rather than the abbreviated P and G (notation):

- I KNOW: *pins* and *gap*
- I WANT: a way of calculating the number of threads (call it *threads*) in terms of *pins* and *gap*.

No pattern has emerged yet, so I must extend the table. Why does each row get longer? It strikes me that a gap of 4 is perfectly possible for 3 pins – why not? Also, what about 2 pins, and even just 1 pin?

FILL IN MORE OF THE TABLE NOW

Why am I doing all this specializing? I WANT to find a pattern in the numbers, but also I WANT to get a feel for what is going on. I noticed as I was filling in the table, that

when *gap* = 1,　　　　　*threads* = 1
when *gap* = *pins*,　　　*threads* = *pins*
when *gap* = *pins*/2,　　*threads* = *gap*
A gap of *gap* and of (*pins* − *gap*) require the same number of threads.
When *gap* is a divisor of *pins*, *threads* = *gap*.

I am led to make two conjectures:

Conjecture 1:
  gap and (*pins* – *gap*) give rise to the same number of threads.

Conjecture 2:
  threads = *gap*, when *gap* is a divisor of *pins*.

Does Conjecture 2 work when *gap* is not a divisor of *pins*? **No!**

  gap = 6, *pins* = 4 requires 2 threads not 6

and

  gap = 4, *pins* = 6 requires 2 threads not 4

I am STUCK! Checking the thread diagram in these cases, it occurs to me that in both cases *threads* is 2, and 2 divides both *pins* and *gap*. I need to try some more complex cases such as

  gap = 6 and *pins* = 9
  gap = 8 and *pins* = 12

and

  gap = 12 and *pins* = 15

TRY SOME NOW

Looking back at my extended table it slowly dawns on me that the number of threads always divides both *pins* and *gap*. AHA! In each case, *threads* is the largest divisor of both *pins* and *gap*. I WONDER if that always works?

Conjecture 3:
  The number of threads is the greatest common divisor of *pins* and *gap*.

  CHECKING for

  gap = 6, *pins* = 8

and

  gap = 8, *pins* = 6

my conjecture seems to work.

  I am now more convinced that it works, but WHY does it work? Will it always work? I WANT an argument to convince me that my CONJECTURE is always right. So, suppose I KNOW the values of *gap* and *pins*. I am still STUCK!

After some time looking at the pins reached by a single thread, and wondering why *threads* should divide both *pins* and *gap*, I realize that I am STUCK! again. Reviewing what I KNOW, I light on the observation that when two divides both *pins* and *gap* I can reach only half the pins. CHECKING cases when three divides both *pins* and *gap*, it seems I can reach only one-third of the pins with a single thread.

AHA! Let me be bold and INTRODUCE something to stand for the greatest common divisor of *pins* and *gap*: why not *gcd*? Now what do I KNOW about *gcd* when I do the threading? Every time I leave the *gap*, what happens in terms of *gcd*? Well, I KNOW that *gcd* divides into *gap*. AHA! Every time I leave the *gap*, I am really jumping around by a multiple of *gcd*. Since *gcd* divides *pins*, I can only ever hope to reach *pins/gcd* of the pins with one thread. Right! This means I must use *gcd* number of threads, as conjectured!

REFLECTING – The greatest common divisor arose spontaneously, as a result of my specializing. However, I was not specializing mindlessly. I was seeking inspiration in the doing of the examples, trying to detect a pattern not just in the number but in the act of looping thread around pins. So *gcd* was the key idea. For me a key moment that stands out is the moment when I decided to use *pins* and *gap* to denote the number of pins and the size of the gap. I could have used *P* and *G*, and would have resorted to them later if a lot of algebra had been involved. By using the words, I avoided having to recall the meaning of *P* and *G*.

While reflecting, another way of expressing the argument came to me. Think of the pins as being set out uniformly round the circle like the hours of a clock, and imagine a single clock-hand pointing to one of the pins. The act of threading can be represented as rotating the clock-hand. Leaving *gap* corresponds to rotating the hand by *gap/pins* of a full revolution. For one thread to go round all the possible pins it can reach means finding the smallest multiple of *gap/pins* which is itself a whole number. *Pins* is one such, but *pins/gcd* is the smallest one, and this means that I need *gcd* threads altogether.

EXTENDING – I can think of many ways to extend the question, but most of them seem difficult. For example:

- How many crossings will one thread make?
- Try replacing a constant gap by a sequence of gaps like 1, 2, 1, 2, . . .
- Try letting the gaps be chosen by the rolls of a die. Then ask about the probability of needing only one thread.

I hope that *Threaded Pins* provided some experience of getting stuck and overcoming it, even if you did not reach a full resolution. There **are** things that you can do when you get stuck. The only way to gain confidence in them is to apply them to get yourself going again after becoming bogged down. Then you will see how effective they are, which in turn will encourage you to tackle harder questions in the future.

There is a story about an undergraduate who arrived late for a lecture and copied down the questions on the board which he took to be the homework. A week or so later he met the professor and complained that the homework seemed rather difficult. In fact he had only been able to resolve two of the questions. The professor then told him that the questions were actually famous unsolved problems! Not **knowing** that the questions were supposed to be difficult made it possible for him to work on the questions without bias. He was not put off by any sense of lack of confidence. The important point is that your attitude may easily affect your possibilities of success.

Here is another question to practise on. You may find it more challenging than the questions in earlier chapters, but if you put into practice what you have learned, you should be able to make progress. **Good luck!**

Leapfrogs

Ten pegs of two colours are laid out in a line of 11 holes as shown. I want to interchange the black and white pegs, but I am only allowed to move pegs into an adjacent empty hole or to jump over one peg into an empty hole. Can I make the interchange?

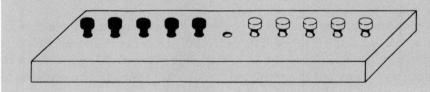

TRY IT NOW

STUCK?

Entry
➤ Have you tried it with coins or pegs?
➤ Have you specialized to fewer pegs?

Attack
➤ What moves block progress? Can you avoid these?

Review
➤ When you have found out how to do it, write down a succinct set of instructions to make the interchange. It's not as easy as it sounds, but well worthwhile. Note any features of the interchange process.
➤ Extend!

I hope that you were not satisfied with the question posed, and that you modified it or posed a new question. For example, you might well have considered other numbers of pegs. More interestingly you might have asked whether the interchange method can be done another way, and what the minimum number of moves is. This is really the most interesting part of *Leapfrogs*.

What is the minimum number of moves?

TRY IT NOW

A resolution

I begin by trying to make the interchange, because I WANT to find out if it is possible, and also because, although I think I understand the rules, I must make sure by trying them out. I decided not to let my pegs go backwards. My first attempts fail and I am STUCK! Perhaps it cannot be done! A strategy which keeps one colour together will not work because I end up with the space where I cannot use it.

What does specialize mean in this context? How about trying it with fewer pegs? With one peg on each side it is easy. However, I will follow my own instructions and record systematically what I do (organization, notation). I shall write B and W for black and white pegs respectively, and leave a space for the empty hole.

	Start	B	W
move	B to the right		B W
leapfrog	W to the left	W B	
move	B to the right	W	B

With two pegs on each side, the strategy that I just used leaves two pegs of the same colour adjacent and I am blocked.

	Start	B B	W W	
move	B to the right	B	B W W	
leapfrog	W to the left	B W B	W	
move	B to the right	B W	B W	
leapfrog	W to the left	B W W B		**Oops!**

AHA! It must be something to do with keeping the colours alternating (conjecture). After several attempts I found how to follow this principle and perform the interchange with two pegs of each colour.

1	B B W W
2	B B W W
3	B W B W
4	B W B W
5	B W W B

6	W B W B
7	W B W B
8	W W B B
9	W W B B

Now I will try it with more pegs and see if the same principle works. I can now make the interchange, though I am still insecure about my methods. I should write it down carefully and CHECK it.

DO SO NOW

The result of writing it down and checking it against examples is that I now see that once I have chosen the first peg to move, my strategy determines a unique peg to move at each stage. To make that last statement true I have to alter my strategy so that at no point do I simply reverse the previous move because several times at the beginning I unwittingly ended up at the starting configuration. I have done enough examples to see why it happens. Now I can ask how many moves it takes.

I need to INTRODUCE a table. In fact I forgot to record the numbers of moves, so I had to go back and count them again!

Number of pegs on each side	Minimum number of moves
1	3
2	8
3	15
4	24
5	35

So the answer to the question of 5 pegs on each side is 35 moves but I would like to know the answer for any number of pegs, say *pegs* on each side. Looking at the pattern of the number of moves leads me to conjecture that the number of moves is always one less than a square. Which square? Why, the square of (*pegs* + 1)! So, 6 pegs should yield (7 × 7) − 1 as the minimum number of moves. CHECK it!

I can now state a generalization:

$$\text{number of moves} = (pegs + 1)^2 - 1$$

where *pegs* is the number of pegs on each side.

I am certainly bothered by the question 'Why?' I want to explain the pattern I have found but how can I find out more? I must specialize by looking more closely at the moves. There seem to be two different kinds: slides and jumps. I shall seek a pattern in each of these. That means going back and counting slides and jumps separately!

Number of pegs on each side	Number of slides	Number of jumps
1	2	1
2	4	4
3	6	9
4	8	16
5	10	25

AHA! Look at that. The number of jumps is the square of the number of pegs on each side and the number of forward moves is equal to the total number of pegs. But why? I am not fully satisfied by the number pattern above (which is most convincing but is still a conjecture) because I WANT to know what the connection is between the rules and the numbers.

It seems to me that the number of slides plus the number of jumps must be equal to the number of spaces that the pegs have to move past. Perhaps I can find this number!

With   B   W   each peg has to move 2 places, giving 4 shifts altogether.

With   B B W W   each peg has to move 3 places, giving a total of 12 shifts altogether.

I have introduced 'shifts' as notation for the total number of spaces the pegs must move.

With   B B B   W W W each peg has to move 4 places, giving a total of 24 shifts altogether.

The pattern is clear enough. If the number of pegs on each side is *pegs* then each peg has to move (*pegs* + 1) places. Thus there are

$$2 \times pegs \times (pegs + 1)$$

shifts. I have resolved and checked that subsidiary question, but what does it have to do with the original?

AHA! How many jumps must I have? Each peg has to go past all the pegs of the opposite colour. Each time that happens, a jump is needed. Thus each white peg has to jump or be jumped by each of the black pegs, so each white peg is involved in *pegs* jumps. Thus there must be *pegs* × *pegs* jumps altogether. That's what I noticed in my last table. Great!

Now what do I KNOW?

Total number of shifts = $2 \times pegs \times (pegs + 1)$
Total number of jumps = $pegs \times pegs$

Now what? Each jump accounts for a shift of 2. AHA!

Total number of shifts = number of slides + 2 × number of jumps

So number of slides = total number of shifts − 2 × number of jumps
$$= 2 \times pegs \times (pegs + 1) - 2 \times pegs$$
$$= 2 \times pegs$$

So now I can find the total number of moves because

Total number of moves = jumps + slides
$$= pegs^2 + 2 \times pegs$$
$$= pegs \times (pegs + 2)$$

REFLECTING – The key ideas were not being satisfied with a number pattern, seeking **why** the conjecture might be true, and breaking up what I WANTED into smaller parts (slides and jumps). The key moment that stands out for me was realizing how often I INTRODUCE notation like shifts without being careful and precise. I shall watch for this in future.

As I review my resolution, I discover that I conjectured the number of moves to be

$$(pegs + 1)^2 - 1$$

but I ended up with

$$pegs \times (pegs + 2)!$$

Since

$$(pegs + 1)^2 - 1 = (pegs + 1) \times (pegs + 1) - 1$$
$$= pegs \times pegs + 2 \times pegs$$

they are the same after all. It is so easy not to notice such details unless time is taken for a thorough review.

On CHECKING, I notice that I have not clearly sorted out why this calculation gives the minimum number of moves. Nor have I thought about the relationship between my strategy and the minimum number of moves. Because the total number of shifts is fixed (pegs do not move backwards) the number of moves will be a minimum when the number of jumps is a maximum. The only way to increase the number of jumps would be to make pegs jump over pegs of the same colour. But I think that the interchange would then fail. When I ask why, I am back into Attack!

Summary

Being STUCK! is a healthy state, because you can learn from it. This will stand you in good stead when you hit harder questions that get you thoroughly bogged down. Recognizing and accepting being STUCK! is not quite so easy or automatic as it sounds. Often you can be stuck, but not be sufficiently aware of it to do any-thing about it.

Once you are aware of being STUCK! do not panic. Relax, accept it and enjoy it, for it is a great opportunity to learn. Specializing is your best friend, elaborated by all of the Chapter 2 advice on Entry. When a new idea

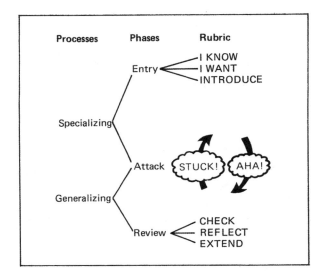

comes and you get unstuck, write down briefly what your idea is. If what you are going to try feels good, why not write AHA!? It makes you feel even better!

In Chapter 10, the following questions might provide enjoyable experiences of being stuck and of overcoming it.

Diagonals of a Rectangle   *Rolling Coins*
Cartesian Chase   *Thirty-one*
Odd Divisors

See Chapter 11 for other curriculum-related questions.

Try Leapfrogs with different numbers of pegs on each side.

4

ATTACK: conjecturing

This is the beginning of a three-chapter attack on the Attack phase. The central activity is conjecturing, and this chapter concentrates on conjecturing WHAT is true. Chapter 5 deals with convincing yourself and others by justifying your conjecture and Chapter 6 deals with what to do when all else fails.

What is conjecturing?

Ask a mathematician what a conjecture is and the reply may well be illustrated by

> ### Goldbach's Conjecture
> Every even number greater than 2 is the sum of two prime numbers. (Note that 1 is not a prime number so 2 has to be excluded.)

A great deal of evidence has been collected which supports Goldbach's conjecture. For example, many millions of even numbers have been tested and every one of them has been found to be the sum of two prime numbers. However, no one has yet proved that **every** even number has this property, and it may not even be true.

A conjecture is a statement which appears reasonable, but whose truth has not been established. In other words, it has not been convincingly justified and yet it is not known to be contradicted by any examples, nor is it known to have any consequences which are false. Goldbach's conjecture is one of the most famous of the many outstanding conjectures in mathematics. Unlike most, it

is easy to state and attempts to justify it have led to a host of subsidiary results and methods. This is typical of an important conjecture.

Not all conjectures have such importance, indeed most are false and are modified almost as soon as they come into being. Yet conjecturing on a small scale lies at the heart of mathematical thinking. It is the process of sensing or guessing that something might be true and investigating its truth. The resolution of *Patchwork* (Chapter 1), for example, contains in an informal way small-scale conjectures that 4, then 3, then 2 colours suffice and that the 'opposite rule', then the 'adjacent rule' will produce proper colourings.

Conjectures like these form the backbone of mathematical thinking. Some property is thought to be true. A conjecture about it often begins as a vague feeling lurking in darkness at the back of the mind. Gradually it is dragged forward by attempting to state it as clearly as possible, so it can be exposed to the strong light of investigation. If it is found to be false it is either modified or abandoned. If it can be convincingly justified, then it takes its place in the series of conjectures and justifications that will eventually make up the resolution. Conjecturing can be pictured as a cyclic process:

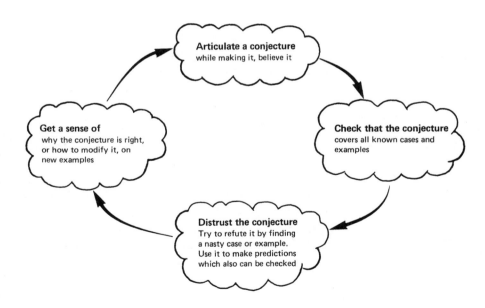

As usual, the best way to appreciate a process is to experience it!

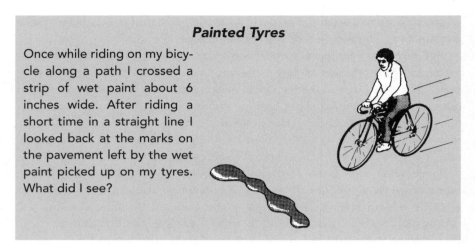

Painted Tyres

Once while riding on my bicycle along a path I crossed a strip of wet paint about 6 inches wide. After riding a short time in a straight line I looked back at the marks on the pavement left by the wet paint picked up on my tyres. What did I see?

CONSIDER IT NOW

STUCK?

➤ The front tyre leaves a series of marks on the pavement. What about the back tyre?

There seem to be two possibilities that people find attractive. One conjecture is that marks appear one tyre-circumference apart. The second conjecture is that two series of marks appear, one from the front tyre and one from the back. This latter conjecture is sometimes qualified by observing that it depends on the distance between the bicycle wheels.

COMMENT ON THE VALIDITY OF THESE CONJECTURES!

Did you notice that having the conjectures written down made it easier to think about them? Half the battle with mathematical thinking is getting into the question sufficiently to get a sense of what might be true, and articulating it as a conjecture. In this case it is a matter of thinking of the possibilities. Having described them succinctly, it is then possible to think about which one seems more plausible, checking each conjecture against **all** known facts.

Did you decide that the two tyres made separate marks? If so, what would happen if two people on unicycles came by a few minutes apart? Altering the conditions often shows what matters. In this case, as long as the wheels have the same radius, they will deposit marks in the same spots.

EXTENSIONS

What if, as is usual, the tyre pressures are not the same and my weight is not uniformly distributed on both wheels?
What if I do not ride in a straight line?

The main point arising from *Painted Tyres* is that in an open-ended question, a conjecture is useful for focusing attention. Articulating a vaguely sensed idea gives the mind something concrete to examine critically. Once articulated, it is important not to believe your conjecture, but this theme will be taken up more strongly in the next chapter. For the moment it is valuable to concentrate on how conjectures arise.

Furniture

A very heavy armchair needs to be moved, but the only possible movement is to rotate it through 90 degrees about any of its corners. Can it be moved so that it is exactly beside its starting position and facing the same way?

TRY IT NOW

STUCK?
Entry
➤ Have you specialized with a cardboard or other model?
➤ Have you found a way of recording the possible moves so that a pattern could emerge?

Attack
➤ Do you think it can be done? Make a conjecture!
➤ Have you asked a more general question – what positions can be reached?
➤ Try using an arrow to represent the way the chair faces, and keep a record of which way the arrow faces after each move.
➤ Are coordinates of any use?
➤ What points can a fixed corner of the chair reach? (Specializing)

TRY IT NOW

It takes only a few minutes of experimenting to have a sense of hopelessness arise inside. When stated out loud as

> I don't believe it's possible

you have a conjecture. Notice the big difference between the earlier vague feeling of hopelessness alternating with fresh hope from each 'but maybe', and the attitude accompanying the assertion, however tempered, 'I don't believe it's possible.' Now there is something to tackle! The conjecture naturally leads on to questions like

> Why can it **not** be done?
> All right, what **can** be done?

The transition to asking what **can** be done is an important aspect of conjecturing because by opening up the original question, generalizing or altering it, a larger pattern may emerge. In this case, following the direction the chair faces, or the progress of each corner in turn, leads to a familiar chessboard pattern. Discussion of this question will be carried further in the next chapter which considers how to justify a conjecture.

Conjecture: backbone of a resolution

The examples in the previous sections and the experience of working on the questions in earlier chapters should have provided plenty of experience in making conjectures. This section is a case study of the role of conjecturing in resolving one question. The question is an interesting one, particularly as it is amenable to many different approaches. Don't read my remarks until you have tried it seriously, and don't be surprised if you come up with a quicker way of doing it. As with *Palindromes* in Chapter 1, confidence with the use of symbols is likely to produce a neater resolution, although not necessarily one whose full implications are understood. The purpose in presenting this case study is to illustrate the way in which conjecturing runs right through the process of thinking mathematically.

> ## *Consecutive Sums*
> Some numbers can be expressed as the sum of a string of consecutive positive numbers. Exactly which numbers have this property? For example, observe that
>
> $$9 = 2 + 3 + 4$$
> $$11 = 5 + 6$$
> $$18 = 3 + 4 + 5 + 6$$

TRY IT NOW

STUCK?
➤ Try lots of examples.
➤ Try changing the question, extending its scope in some way.
➤ Be systematic in your specializing and try several different systems.
➤ Look for patterns.

A resolution
Entry
Begin by specializing. Two systematic approaches come to mind. Either take each number in turn and try to express it as a sum of consecutive numbers or be

systematic and take sets of two, then three, then four consecutive numbers and find the sums. For the moment, choose the first alternative.

Attack

$1 = 0 + 1$ Is 0 allowed?

No, numbers have to be positive.

$2 = ?$ Can't be done.

$3 = 1 + 2$

$4 = ?$ Can't be done.

Conjecture 1:

Even numbers are not the sum of consecutive numbers.

Continue the attack with more specializing:

$5 = 2 + 3$

$6 = 1 + 2 + 3$

Conjecture 1 is disproved. Continue specializing:

$7 = 3 + 4$

$8 = ?$ Can't be done.

Conjecture 2:

Powers of two are not the sum of consecutive numbers.

The evidence for Conjecture 2 is rather thin, though since $1 = 2^0$ it copes nicely with 1. That is nice, because I had forgotten him. Now I predict that $16 = 2^4$ will be in trouble. More specializing is indicated, up to 16 at least. While accumulating all this data, there are a lot of other patterns emerging, concerned with sums of two, three and four numbers. These patterns should be written down as AHA!s or conjectures, even if at this stage they are not checked thoroughly. Some of them may contain important observations which will be useful as the resolution progresses.

DO THIS NOW, IF YOU HAVE NOT ALREADY

REFLECT: Pause now and notice how the conjecturing process is already well under way. Conjecturing is arising automatically by carrying out the familiar processes of specializing and generalizing. Specializing gives a feel for what is going on; detecting some underlying pattern (generalizing) and articulating it produces a conjecture which can then be examined, challenged and modified. In this case, more specializing has supported Conjecture 2.

The conjecturing process so far looks something like the diagram on the next page. Conjecture 2 has been round a full cycle, and further examples seem to confirm it. Before thinking in more general terms about why it might or might not be true, notice that it can be substantially improved by articulating a second feature to which the specializing has pointed.

Conjecture 3:

(i) **Powers of two cannot be expressed as sums of consecutive positive numbers.**

(ii) **All other numbers are sums of consecutive positive numbers.**

To see whether Conjecture 2 is true for all numbers, we really have two subsidiary questions to answer:

1 How can any number that is **not** a power of two actually be written as the sum of consecutive numbers?

2 Why can a power of two **not** be expressed as a sum of consecutive numbers?

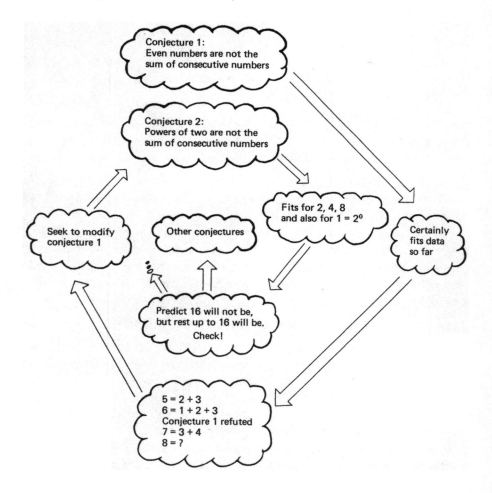

These two questions lead to a fundamental issue: what distinguishes a power of two from other numbers? How does this distinguishing property relate to the property of being a sum of consecutive numbers? I KNOW that a power of two has no prime factors other than two, by its very definition. All its factors other than one are therefore even numbers. For example, the factors of 16 are:

16, 8, 4, 2 and 1

which are all even except 1, whereas the factors of 22 are:

22, 11, 2 and 1

and there is an odd factor other than 1. I am not yet sure that this is relevant, but it is worth recording as a remark. It actually has a status somewhere between a conjecture and a fact, depending on how sceptical I want to be in my final resolution.

Conjecture 4:
Every number which is not a power of two has an odd factor other than 1.

I have recorded it as a conjecture because I do not want to spend time now justifying it. It feels right and, since it is written down clearly, I will check it in my Review. My mathematical experience reinforces my observation with a strong sense of its verifiability and I am confident enough to proceed without diverting my concentration from the main thread of the investigation. However, it helps to record such points so that they will be checked again later when I am more calm and less caught up in the flow of ideas.

How does the existence of an odd factor enable me to express the number as a sum of consecutive numbers? Specialize by examining numbers with odd factors, say multiples of 3 and 5.

$$
\begin{aligned}
3 &= 1 + 2 \\
6 &= 1 + 2 + 3 \\
9 &= 2 + 3 + 4 \\
12 &= 3 + 4 + 5 \\
15 &= 4 + 5 + 6 \\
5 &= 2 + 3 \\
10 &= 1 + 2 + 3 + 4 \\
15 &= 1 + 2 + 3 + 4 + 5 \\
20 &= 2 + 3 + 4 + 5 + 6 \\
25 &= 3 + 4 + 5 + 6 + 7
\end{aligned}
$$

A clear pattern (and hence a conjecture) is emerging here as larger multiples of 3 (and 5 respectively) can be obtained by taking the sum of 3 (or 5) consecutive numbers with larger starting numbers. It is usually worthwhile deliberately trying to articulate a conjecture in a situation like this because it forces you to clarify your feeling for what might be going on and provides something concrete to test. However, especially with a conjecture that arises during exploration, there is no need to be excessively formal. The important thing is to begin to capture the ideas involved. A first attempt might look like this:

Conjecture 5:
A number which has an odd factor can be written as the sum of consecutive numbers. Usually the odd factor will be the same as the number of terms.

It took five attempts to word Conjecture 5 so that it made sense! I have deliberately inserted the 'usually' because I am not ready to be precise, but I do not want to be put off by those mavericks at the start. In order to do anything with Conjecture 5 I am going to have to manipulate some odd factors. It would be best to INTRODUCE some symbols. Notice that odd numbers are characterized by the shape $2K + 1$, where K is an integer.

Conjecture 5A:
A number N which has an odd factor $2K + 1$ is usually the sum of $2K + 1$ consecutive numbers.

This fits all the data collected so far. The sensible thing is to try to see what is going on while checking it systematically on new examples.

Multiples of 3

$$3 \times 2 = \quad 1 \quad + 2 + \quad 3$$
$$3 \times 3 = \quad 2 \quad + 3 + \quad 4$$
$$3 \times 4 = \quad 3 \quad + 4 + \quad 5$$
$$3 \times F = (F - 1) + F + (F + 1)$$

Multiples of 5

$$5 \times 3 = \quad 1 \quad + \quad 2 \quad + 3 + \quad 4 \quad + \quad 5$$
$$5 \times 4 = \quad 2 \quad + \quad 3 \quad + 4 + \quad 5 \quad + \quad 6$$
$$5 \times 5 = \quad 3 \quad + \quad 4 \quad + 5 + \quad 6 \quad + \quad 7$$
$$5 \times F = (F - 2) + (F - 1) + F + (F + 1) + (F + 2)$$

Looks good! I am beginning to get the idea that I get a multiple of the middle number. Thus if

$$N = (2K + 1) \times F$$

then N is the sum of $2K + 1$ consecutive numbers, the middle one being F.

Conjecture 6:

TRY expressing N, which is $F \times (2K + 1)$, as a sum of F_3

$$
\begin{array}{lcl}
\qquad\qquad F & = & F \\
(F - 1) + (F + 1) & = & 2F \\
(F - 2) \qquad + \qquad (F + 2) & = & 2F \\
\cdots & & \\
(F - K) \qquad + \qquad\qquad (F + K) & = & 2F
\end{array}
$$

There are $K + 1$ equations here, so the sum of all the left-hand sides is the sum of all the right-hand sides which is

$$(K \times 2F) + F$$

that is

$$(2K + 1) \times F$$

AHA! It works!

CHECK! There are $2K + 1$ consecutive terms. Oh dear! Are they all positive? Only if F is large enough. Let me see some examples of that, following the pattern of Conjecture 6.

$$
\begin{array}{ll}
F = 1: 3 \times 1 = & (1 - 1) + 1 + (1 + 1) \\
\qquad\qquad = & \quad 0 \quad + 1 + \quad 2
\end{array}
$$

$$
\begin{array}{ll}
F = 1: 5 \times 1 = & (1 - 2) + (1 - 1) + 1 + (1 + 1) + (1 + 2) \\
\qquad\qquad = & \quad -1 \ + \ \ 0 \ \ + 1 + \ \ 2 \ \ + \ \ 3
\end{array}
$$

$$
\begin{array}{ll}
F = 2: 5 \times 2 = & (2 - 2) + (2 - 1) + 2 + (2 + 1) + (2 + 2) \\
\qquad\qquad & \quad 0 \ \ + \ \ 1 \ \ + 2 + \ \ 3 \ \ + \ \ 4
\end{array}
$$

$$
\begin{array}{l}
F = 1: 7 \times 1 = (1 - 3) + (1 - 2) + (1 - 1) + 1 + (1 + 1) + (1 + 2) + (1 + 3) \\
\qquad\qquad\ = \ \ -2 + \ \ -1 + \ \ \ 0 \ \ + 1 + \ \ 2 \ \ + \ \ 3 \ \ + \ \ 4
\end{array}
$$

$$
\begin{array}{l}
F = 2: 7 \times 2 = (2 - 3) + (2 - 2) + (2 - 1) + 2 + (2 + 1) + (2 + 2) + (2 + 3) \\
\qquad\qquad\ = \ \ -1 + \ \ 0 \ \ + \ \ 1 \ \ + 2 + \ \ 3 \ \ + \ \ 4 \ \ + \ \ 5
\end{array}
$$

$$
\begin{array}{l}
F = 3: 7 \times 3 = (3 - 3) + (3 - 2) + (3 - 1) + 3 + (3 + 1) + (3 + 2) + (3 + 3) \\
\qquad\qquad\ = \ \ \ 0 \ \ + \ \ 1 \ \ + \ \ 2 \ \ + 3 + \ \ 4 \ \ + \ \ 5 \ \ + \ \ 6
\end{array}
$$

AHA! I can always forget the zero, and the negative terms are always counteracted by the positives. Thus

$$
\begin{array}{lll}
3 \times 1 = & 0 + 1 + 2 & = 1 + 2 \\
5 \times 1 = & -1 + 0 + 1 + 2 + 3 & = \quad 2 + 3 \\
5 \times 2 = & 0 + 1 + 2 + 3 + 4 & = 1 + 2 + 3 + 4 \\
7 \times 1 = -2 + & -1 + 0 + 1 + 2 + 3 + 4 & = \qquad\quad 3 + 4 \\
7 \times 2 = & -1 + 0 + 1 + 2 + 3 + 4 + 5 & = \quad 2 + 3 + 4 + 5 \\
7 \times 3 = & 0 + 1 + 2 + 3 + 4 + 5 + 6 = 1 + 2 + 3 + 4 + 5 + 6
\end{array}
$$

Perhaps I can use this to deal with the 'usually' of Conjecture 5. Recap: I KNOW that any number N with an odd factor $2K + 1$ can be written as the sum of $2K + 1$ consecutive numbers. But some of them may be negative. I WANT to show that any number N with an odd factor can be written as the sum of two or more consecutive positive numbers.

Pondering on I WANT and I KNOW for a while, I suddenly realized that I had actually finished! All I have to do is counteract the negative terms by the corresponding positive terms. There must be more positive terms than negative ones since the total sum is positive!

CHECK! What if this counteracting of positive and negative leaves me with just a single term. Oh dear! Could this happen? Specialize:

$$-2 + -1 + 0 + 1 + 2 + 3$$

has 6 terms. It is the presence of 0 which does it. To end up after the counteracting with just one term I would need to have had an even number of terms all told; but I have $2K + 1$ terms which is always odd. This idea generalizes:

Conjecture 7:
Starting with an odd number of terms including 0 the counteracting process always leaves me with an even number of consecutive positive numbers.

I am now satisfied that every number divisible by an odd factor other than 1 **can** be written as the sum of consecutive positive numbers. Flushed with success, I pause to review my work.

REVIEW: I have answered subsidiary question 1, but not question 2: why cannot a power of two be expressed as a sum of consecutive positive numbers?

Surely my work so far contains the answer? Let me see. Suppose the number N **can** be written as the sum of consecutive positive integers. Look at

$$7 = 3 + 4 \text{ and } 5 = 2 + 3$$

Now earlier I got these as

$$7 = -2 + -1 + 0 + 1 + 2 + 3 + 4$$

and

$$5 = -1 + 0 + 1 + 2 + 3$$

AHA! Why not use the counteracting idea again? Take any sum of consecutive positive integers. Extend them downwards to zero and beyond so that the negative ones counteract the additional positive ones. Now I WANT to show that if N can be written as the sum of consecutive positive numbers, then it **must** have an odd factor. AHA! It all depends whether there are an odd number of terms in the sum. I detect two cases:

(i) N has been expressed as the sum of an odd number of consecutive positive numbers.

(ii) *N* has been expressed as the sum of an even number of consecutive positive numbers.

Now case (i) looks to me like Conjecture 6 backwards. I should be able to deduce that in case (i) *N* has an odd factor. AHA! case (ii) looks like the counteracting idea of conjecture 7.

So I think I am ready to write down an argument. Several refinements later, it comes out as:

Conjecture 8:
If *N* can be written as the sum of an odd number of consecutive (not necessarily positive) numbers, it has an odd factor.

Argument:
Using the format of Conjecture 6, suppose there are $2K + 1$ terms. Then the terms can be grouped around the central term *F*, as

$$F$$
$$(F - 1) + (F + 2)$$
$$(F - 2) \qquad + \qquad (F + 2)$$
$$\cdots$$
$$(F - K) \qquad + \qquad (F + K)$$

$$\text{Thus } N = F \times (2K + 1)$$
Thus *N* has an odd factor.

Now I am ready to turn back to Conjecture 2:

Conjecture 2:
Powers of two are not the sum of consecutive positive numbers.

Argument:
Any number *N* which is a sum of consecutive positive numbers is either the sum of an odd number of terms or an even number of terms. If the number of terms is odd, the argument for Conjecture 8 shows than *N* must have an odd factor and so cannot be a power of two. If the number of terms is even then augment these down to 0 and beyond, to produce an odd number of consecutive terms whose sum is still *N*. Again, *N* must have an odd factor by Conjecture 8, and so cannot be a power of two.

Now I think I have finished.

REVIEW: Checking over the details everything seems in order. A final writing out of the arguments might make it crisper, though once the trail of conjectures is lost, the resolution becomes rather sterile. There were a number of key ideas. Most salient was introducing or permitting negative numbers even though I knew I had to get rid of them eventually. They allowed me to see Conjecture 6 and subsequently provided the basis for my resolution. There were several key moments for me. The need for, and value of, moving to symbols N, $2K + 1$ and F remains strongly with me. They gave me something specific yet general to manipulate. I note that using one as an odd factor expresses any number as a sum of only one positive 'consecutive' number, including the powers of two.

Introducing symbols is a powerful technique for increasing the amount of information given by a conjecture, yet keeping it readable. Each symbol used must be clearly specified and great care should be taken to ensure consistency of meaning. It is always a good idea to check any symbolic statement on numerical cases, before any work is done using the new statement. This is also the technique to adopt if you feel ill at ease with symbols. Keep referring back to numerical instances and interpret the statement in those cases.

I was struck also by the quick flow of conjectures and the pleasure they gave me. Being aware of the spiral nature of conjecturing prompts me to give more care to conjectures than I once did. When I am stuck I can go back and look for where I departed from the spiral by failing to test and to disbelieve my conjectures.

Conjectures are like butterflies. When one flutters by, there are usually many more close behind. As each one comes by, it attracts attention away from the last one, and so it is easy to lose track. Once they start coming in profusion it is wise to try to pin them down in a few words so that you can return to them later. You will find that, like butterflies, conjectures are not easy to capture. It may take several attempts. In the process of trying, your mind is focused and the conjecture takes shape instead of remaining imprecise. Real thinking about the conjectures is then possible.

It is important also to recognize when a statement has the status of a conjecture and when it has been convincingly justified. Dogmatic belief in a forcefully made statement is not thinking! It is useful to treat every statement as a conjecture which needs checking and justifying. In *Consecutive Sums* I was careful to mark each conjecture, and to go back and provide arguments to justify the key ones. How to justify a conjecture is the subject of Chapter 5.

Finally, EXTENDING the resolution, I accumulated a lot of data and I noticed that

$$9 = 4 + 5 = 2 + 3 + 4$$
$$15 = 7 + 8 = 4 + 5 + 6 = 1 + 2 + 3 + 4 + 5$$

I wonder if I could predict in how many ways a given number is the sum of consecutive positive integers? See *Consecutive Sums* in Chapter 10.

How do conjectures arise?

My resolution to *Consecutive Sums* illustrates the process of conjecturing as it often happens in mathematical thinking. But where do conjectures come from and how do they arise? The most important thing to realize is that you need confidence and/or bravado. If you are timidly tentative, never willing to try something out then reject it or modify it, then you are unlikely to realize your potential. However, confidence is not gained by saying to yourself

I will be confident!

It comes from past successes and from the release of inner tensions that accompany feeling STUCK. In order to make a step towards such a release, I highly recommend augmenting your RUBRIC writing (for the last time) to writing down the briefest of notes to yourself whenever an idea comes. Little things like

TRY . . .
MAYBE . . .
BUT WHY . . . ?

are very helpful in two ways. Firstly, they focus your attention on the idea so that it doesn't get swallowed by your next thought, and, secondly, they help you to recall what it was you thought you were doing. TRY . . . and MAYBE . . . start off as little notes to yourself. In time you will find that they are turning into full blown conjectures! So, if you are tentative and unsure when working on a question, use TRY . . . , MAYBE . . . and BUT WHY . . .? (and specialize with concrete examples that do give you confidence!).

Conjectures seem to be spawned by two main activities. Specializing, probably the most common source, has already been discussed. The other method is by analogy, which is a form of generalizing.

It can happen that while you are investigating one situation you will suddenly be struck by a similarity with some previously tackled question. Sometimes the similarity will be exact, the two questions being virtually identical but dressed in different clothes. Frequently the similarity will be only partial, but nevertheless very helpful for suggesting conjectures and approaches. Examples of this process at work are hard to give as they are dependent on recent experience and on personal ways of looking at and thinking about questions. For me, an example occurred when, some time after looking at *Consecutive Sums* I came across this question:

Square Differences
Which numbers can be expressed as the difference of two perfect squares?

TRY IT NOW

STUCK?
➤ Specialize: do not give up easily!
➤ Be systematic. Recall that in *Consecutive Sums* there are two ways to generate examples: starting with a number, and starting with a consecutive sum.
➤ It may help to know that every difference of two squares can be factored.

I began by trying to express 1, 2, 3, . . . in turn as the difference of two squares, but with no luck at all. Then the double approach of *Consecutive Sums* suggested being systematic in an analogous way. Thus

$$2^2 - 1^2 \quad 3^2 - 1^2 \quad 4^2 - 1^2 \dots$$
$$3^2 - 2^2 \quad 4^2 - 2^2$$
$$4^2 - 3^2$$

suggested a connection with *Consecutive Sums* and so I was reminded of the idea of factors. A conjecture soon arose, involving numbers that are twice an odd factor. My algebraic experience then took over and I used the standard factoring of

$$N^2 - M^2$$

as $(N - M)(N + M)$ to reach a resolution which CHECKED with my examples.
A friend of mine also found that *Square Differences* was analogous to *Consecutive Sums*, although in a different way. She had resolved *Consecutive Sums* by expressing the sum of the numbers from $N + 1$ to M as

$$(N - M)(N + M + 1)/2$$

and noticing that one of $N - M$ or $N + M + 1$ must be even and the other odd. She applied the same key idea to *Square Differences*, noticing that $N - M$ and

$N + M$ are both even or both odd and followed an analogous path to a resolution. In both cases, the analogy with *Consecutive Sums* turned out in the end to be only partial, but analogies can be exact, often in surprising ways. For example, a question very closely analogous to *Threaded Pins* (Chapter 3) is posed in a later chapter. You might like to keep an eye open for it! The following game is identical to a well-known children's game. Can you see the analogy?

Fifteen

Nine counters marked with the digits 1 to 9 are placed on the table. Two players alternately take one counter from the table. The winner is the first player to obtain, amongst his or her counters, three with the sum of exactly 15.

TRY IT NOW

STUCK?

➤ Play the game enough times to discover the best first move.
➤ Does the sum 15, in the context of the numbers 1 to 9 ring a bell?
➤ What sets of numbers sum to 15?
➤ How many times does each number occur in these sets?
➤ Can you arrange the sets so that their overlap is displayed?

Discovering pattern

The process of conjecturing hinges on being able to recognize a pattern or an analogy, in other words, on being able to make a generalization. Finding the pattern may ultimately be a creative act beyond your direct control but, as with all creative processes, a great deal of groundwork can be done to prepare for an insight. Further specializing is the obvious suggestion. This provides more information and another opportunity to get the feel of the situation as it is worked through once more. Reorganizing information that has already been obtained is another powerful tool. This may be simply a change in layout or it may involve reorganization of your thinking as well.

For example, in *Consecutive Sums* the idea of rearranging

$$3 \times 1 = 0 + 1 + 2 \text{as } 1 + 2$$
$$5 \times 1 = -1 + 0 + 1 + 2 + 3 \text{ as } 2 + 3$$
$$\text{etc.}$$

was crucial for making the reverse step of augmenting sums of an even number of consecutive terms to get an equivalent odd number of consecutive terms.

Generalizing involves focusing on certain aspects common to many examples, and ignoring other aspects. Conjecture 5 in *Consecutive Sums* is an excellent example because of the word 'usually'. The feature which attracted me was not common to every example! The point is that it is not enough to do a lot of examples and then sit back and ask for what is common. Being creative requires becoming thoroughly involved in and imbued with the examples so that they almost begin to speak to you. Then the moment of insight which develops into a valid conjecture brings great pleasure, indeed it sustains you through the longer periods of frustration at wrong conjectures and unseen patterns.

Powers of generalization in mathematics can be increased through practice and thought-provoking exposure to open-ended questions. The two main ways are:

- by developing an expectation of pattern and being prepared to carry out an active search for it;
- by building up mathematical knowledge and experience.

One of the most pleasing and satisfying aspects of mathematics is the rich abundance of patterns that are found in all its branches. Expecting to find regularity in the results of a mathematical investigation is a feeling that grows as one is exposed to mathematical thinking and this predisposes you to discovering and recognizing patterns. In *Consecutive Sums* I was so convinced that a pattern must be present that I was willing to ignore aspects which failed to fit. Success with *Fifteen* depended on a sense that there must be a pattern driving me on until I recalled the similarities with magic squares.

Knowledge of mathematics also helps substantially in finding patterns. Firstly, as particular patterns become more familiar, they also become easier to recognize even when disguised, so that someone well acquainted with the square numbers will probably find it easy to recognize

$$2, 8, 18, 32, 50, 72, \ldots$$

or

$$3, 8, 24, 35, 48, 63, \ldots$$

Secondly, there are certain standard techniques such as examining the difference between successive terms of a number sequence, which can be used to work out what a pattern might be. Thirdly, studying particular branches of mathematics develops an awareness of those functions which are likely to be important in associated situations. For example, *Threaded Pins* (Chapter 3) reminded me of modular arithmetic, where I know that the greatest common divisor is an important idea, so I was open to it. Someone who has never before met the idea of greatest common divisor would probably find *Threaded Pins* much more difficult. So, extending our knowledge extends the field on which your thinking can work. Of course, experience does not always lead in a useful direction because it can produce firmly established opinions. Once you are fixated on an approach, it is hard to change. Part of the art of conjecturing is to be open to new interpretations which arise unexpectedly in what otherwise might seem a familiar context. This topic of fixation is discussed more fully later, in both Chapters 5 and 6.

Even an expectation of pattern, especially of simple pattern, may be misleading. There are some situations where patterns and generalizations are not at all evident and others where the patterns are actually more complicated than one might at first suspect. Here is a good question to try which introduces a note of caution into making generalizations too freely.

Circle and Spots

Place *N* spots around a circle and join each pair of spots by straight lines. What is the greatest number of regions into which the circle can be divided by this means?

For example, when there are 4 spots, 8 regions is the maximum possible number of regions (in this case, 8 is also the minimum).

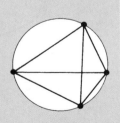

TRY IT NOW

STUCK?

➤ Again, try several examples, carefully.
➤ How can you be sure you have the maximum number of regions?

➤ Do the numbers make a pattern?
➤ CHECK your conjectures on more examples.
➤ How many times do the lines joining the spots cross?
➤ Be sceptical!

Trying the first five cases usually convinces people that S spots will produce 2^{s-1} regions, and so 6 spots will produce 32 regions. People often search the diagram many times for the lost 32nd region before accepting that the obvious pattern leads to a false conjecture. A question like this is a very valuable one, because the surprise it gives remains with you, guarding against over-confidence.

Summary

Conjecturing is the recognizing of a burgeoning generalization. Once conjectures begin to flow, they tend to come like a cloud of butterflies, elusively flying off when approached. It is wise at this time to try to capture some of them and to recall the cyclic process (see the diagram below).

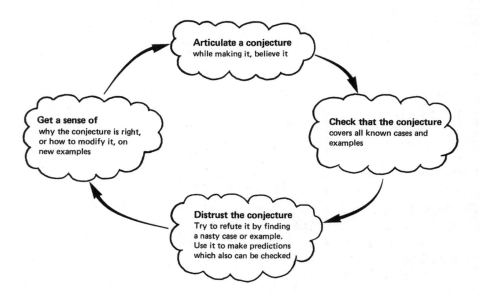

Articulating, testing and modifying conjectures form the backbone of a resolution.

Because conjecturing involves the creative act of generalizing, it is not enough to accumulate examples systematically and expect a pattern to jump out. It requires getting fully involved and imbued with the question. Specializing may have to be reorganized and analogies may have to be explored. Practice and study of mathematical techniques are the best ways to extend your range of possibilities. Finally, conjectures are always suspect. Remember *Circle and Spots*!

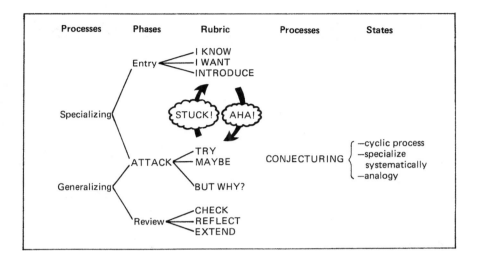

In Chapter 10 the following questions offer similar experiences of conjecturing:

Sums of Squares   *More Furniture*
More Consecutive Sums   *Nullarbor Plain*
Fare is Fair

See Chapter 11 for other curriculum-related questions.

5

ATTACK: justifying and convincing

This chapter tackles two different activities: seeking why and explaining why. Seeking why involves getting a sense of some underlying reason for the truth of your conjecture. Explaining why involves convincing yourself and, more importantly, convincing others that you can justify your arguments. Explaining why is largely based on the idea of mathematical structure, an important notion that lies behind attempts to explain why something might be true, and it is a development of conjecturing.

Structure

During the resolution of questions in the first three chapters, the process of seeking WHAT was true and articulating it as a conjecture was followed swiftly by the process of seeking WHY it was true (or in some cases false!). For example, during *Chessboard Squares* (Chapter 1) the conjectures that arose were quickly justified by simple counting arguments. However, it is often the case that conjecturing WHAT is much easier than seeing WHY. Being able to convince others is often more difficult still. Two extreme examples in which the WHY remains obscure despite much investigation are *Goldbach's Conjecture* (Chapter 4), and the following old chestnut.

Iterates

Choose any integer.

- If your integer is even, divide it by two.
- If your integer is odd, multiply by three, add one and divide by two.

Keep going with your new number! Will you eventually reach one?

DO NOT SPEND LONG ON THIS!

It is known that you will eventually reach one if you start with any number less than 5 billion billion, but little else is known. As with *Goldbach's Conjecture*, a great deal of specializing has already been done and most people are convinced that they are reasonable conjectures. However, no one has so far been able to provide a convincing argument that stands up to scrutiny. What is needed is not simply examples, but some reason, some underlying pattern or structure around which to frame an argument.

Mathematicians have spent a lot of time trying to clarify what they mean by structure; indeed the body of mathematical knowledge could be seen as the current understanding of what structure means. It would be presumptuous to try to give a general definition of what is meant by structure, but by means of some examples you should be able to get a sense of it.

Matches 1

How many matchsticks are required to make 14 squares in a row, the side of each square being the length of a match, as in the following sequence?

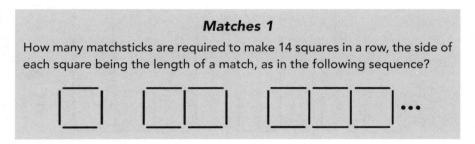

TRY IT NOW

The most obvious thing to do is to count the number of matches in each member of the sequence given (systematic specialization), and then to look for a pattern in the numbers. It takes no great insight to observe that the numbers

4, 7, 10, 13, . . .

are rising by 3 each time. So the conjecture is clear, not only that the 14th member requires 43 matches, but more generally that the Nth member requires $3N + 1$ matches. To justify this statement and to convince a sceptic, you must show why

rising by 3

captures what is going on in the matchsticks.

This example is quite transparent because the matches can be grouped as

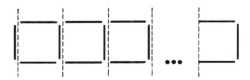

which shows that the Nth member of the sequence requires 1 match to start with, and then N further groups of 3 matches, making $3N + 1$ altogether. This constitutes a convincing argument because it links the conjectured formula (what we WANT) with the **structure** of the matchstick configuration (what we KNOW).

Do not be lulled by the simplicity of this example, however, because very often people notice a number pattern and then mistake their conjectured pattern for a fully justified solution. Try this next one.

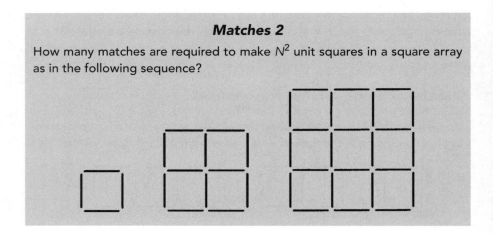

Matches 2

How many matches are required to make N^2 unit squares in a square array as in the following sequence?

TRY IT NOW

STUCK?

➤ Specialize systematically.
➤ Count the matches!
➤ Do the next example. Look for a pattern.
➤ How did you count the matches? Generalize!
➤ Try other systematic methods of counting.

If you counted the matches in each configuration, you probably did not recognize the numbers that came out. However, by looking at HOW you counted the matches, you can generalize your method to count the matches in the Nth configuration. The Nth one has N^2 little squares.

There are $N + 1$ rows   of N horizontal   matches
and $N + 1$ columns   of N vertical   matches

giving $2N(N + 1)$ matches altogether. In this case a conjecture based solely on the number sequence

4, 12, 24, . . .

is pretty hard, but the structure of the matchstick configurations is quite clear.

The underlying structure here, as I have exposed it, consists of what is common between the match pattern (the horizontal and vertical rows) and the number pattern generated by the formula $2N(N + 1)$. The structure itself is hard to speak about, but the matches and the formula are two manifestations of it. The formula is justified convincingly by exposing the link between the method of counting the matches and the method of building up the formula. Whereas in *Matches 1* the justification is based on how to pass from one configuration to the next, in *Matches 2* it is based on a systematic analysis of a single configuration. These two approaches are very common: the iterative or recursive development of one configuration in terms of earlier members of a sequence, and a direct attack on a general configuration.

In both of the *Matches* examples, the structure was captured by a number pattern, but this is not always the case. Structure of quite a different sort arose in my resolution of *Furniture* (Chapter 4). Using an arrow to indicate the way the chair was facing, following the possible movements of the chair, and recording its position with the arrow, led to a chessboard kind of pattern with horizontal and vertical arrows alternating.

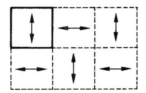

All 90 degree rotations about a corner preserve this chessboard pattern, so the chair cannot end up beside its starting place **and** remain facing in the same direction.

While REVIEWING my work I wondered if another approach might be possible, independent of the chessboard. Introducing coordinates, and tracing the progress of one corner of the chair, chessboard in mind, I noticed that no matter where that corner moved, the sums of the coordinates were always even or always odd depending on which corner I followed. This observation was then turned into a more precise and convincing argument as follows:

Argument:
Suppose one corner of the chair is at (a, b). The only possible positions it can occupy after one 90 degree rotation are

(a, b),	$(a + 1, b + 1)$,	$(a + 2, b)$,
$(a - 1, b - 1)$,	$(a, b - 2)$, and	$(a + 1, b - 1)$.

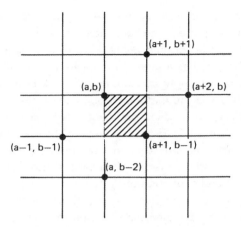

These all preserve the evenness or oddness of the sum of the coordinates. However, the question asks if it is possible to get the chair into a position which reverses the evenness–oddness of the coordinate sum, and that is impossible.

Another example of non-numerical structure appeared in *Fifteen* (Chapter 4) where the magic square captures all the information needed to play the game, and turns the number work into positional play.

4	9	2
3	5	7
8	1	6

The examples in this section give only a small sample of the wide range of ideas covered by the notion of mathematical structure. More sophisticated examples will come with continued exposure to mathematical ideas. The important thing to remember is that a conjecture is an informed guess about a possible pattern or regularity which might explain WHAT is puzzling in a particular question. Once formulated, the conjecture is investigated to see whether it must be modified or whether it can be convincingly justified. This is done by seeking WHY. The nature of I WANT changes, from trying to articulate what is true, to trying to see why the conjecture might be justified, that is,

> from seeking WHAT
> to seeking WHY

The answer to WHY is a structure which links what you KNOW to what you have conjectured. Your argument will be an exposition of that link.

Seeking structural links

This section looks in more detail at the way structure is used to justify a conjecture and to explain convincingly why it is true. *Matches 1* and *Matches 2* illustrate the fact that when you are trying to explain WHY, to justify a conjecture about WHAT is true, there are generally two sources of patterns. One source is the original data, in these cases matchstick configurations, which constitute what you KNOW. The second source is in your conjecture, which is what you WANT to justify. To resolve a question satisfactorily means to find and state clearly a relationship between an underlying pattern in what you KNOW and what you WANT. Sometimes a sense of the common pattern comes directly from what you WANT, as in the

4, 7, 10, 13, . . .

sequence of *Matches 1*, and sometimes it comes directly from what you KNOW, as in *Matches 2*. More frequently it emerges from an interplay back and forth between the two. The common pattern that links KNOW and WANT is the structure. Articulating the link is the basis of a justification.

The *Matches* examples have been intentionally simple, but the principle remains the same in any question. For example, in *Palindromes* (Chapter 1), systematic specializing led to the pattern that successive palindromes differ by 110 or by 11. This pattern was then related to what I KNEW about palindromes by the observation that to pass from one palindrome to the next, you either

increase both the tens and the hundreds digit by one
(add 110)

or

increase the units and thousands digits by one, and decrease the tens
and hundreds by nine
(add $1001 - 990 = 11$)

Similarly in *Patchwork* (Chapter 1), when I specialized systematically I was led to the observation that two colours always sufficed (what I WANT). This conjecture was justified by means of an algorithm in terms of adding lines one at a time and modifying the colouring each time. The algorithm depended on two structural features of patchworks (what I KNOW): that they can be formed by adding lines one at a time, and that when a new line is added, each old region either remains intact or is divided in two. Notice that both *Patchwork* and *Palindromes* required more than just systematic specializing. It was necessary to discern some pattern (what I WANT) and to relate this to the underlying structure (what I KNOW).

Circle and Spots (page 75) gives an example of an obvious pattern in the numbers **not** being reflected in the original data. It is a natural conjecture that the sequence

1, 2, 4, 8, . . .

will continue as powers of 2, but powers of 2 are not involved in the structure of the regions. Finding a pattern in the first few numbers is not enough to resolve the question, and this illustrates the point that it is not enough to formulate a plausible conjecture. You must link it back to the structure of the original data. The next question illustrates how structural links are often found in questions based on counting.

Bee Genealogy

Male bees hatch from unfertilized eggs and so have a mother but no father. Female bees hatch from fertilized eggs. How many ancestors does a male bee have in the twelfth generation back? How many of these are males?

TRY IT NOW

STUCK?

Entry
➤ Draw a diagram or family tree.
➤ Don't draw all 12 generations!

Attack
➤ Look for a pattern in the numbers, and in the diagram.
➤ Have you justified your conjecture? You need a direct link between how the numbers grow, and how the generations grow.

Bee Genealogy is interesting because people familiar with Fibonacci numbers

   1, 1, 2, 3, 5, 8, . . .

in which each term is the sum of the two preceding terms, quickly recognize them in this context, and often think that spotting the pattern solves the question. However, the Fibonacci pattern remains a conjecture until it has been linked directly to the data, in this case the reproductive vagaries of bees. The link is provided by the formula

$$
\begin{bmatrix} \text{number of bee ancestors} \\ N + 2 \\ \text{generations} \\ \text{back} \end{bmatrix} = \begin{bmatrix} \text{number of bee ancestors} \\ N + 1 \\ \text{generations} \\ \text{back} \end{bmatrix} + \begin{bmatrix} \text{number of bee ancestors} \\ N \\ \text{generations} \\ \text{back} \end{bmatrix}
$$

TRY JUSTIFYING THIS FORMULA

Once the structural correspondence is articulated, finding the required answer for 12 generations back is a straightforward computation. You might have conjectured an 'answer' by using the Fibonacci pattern, but you can't be sure of your answer until the link has been forged (recall *Circle and Spots!*).

The structural features of questions based on counting are almost always the same as in *Bee Genealogy*. The pattern in things being counted is mirrored by the pattern in the numbers that arise. Noticing the pattern in the numbers often suggests patterns to look for in the objects themselves, sometimes by reference to another counting problem which gives rise to the same numbers. The structure featured in the next question is more complicated, but again the patterns in what is KNOWN reflect patterns in the WANTED numbers.

Square Dissection

A number N is called 'nice' if a square can be dissected into N non-overlapping squares. What numbers are 'nice'?

TRY IT NOW

STUCK?

Entry

➤ What does dissected mean?

➤ Try simple cases.

Attack

➤ Do enough special cases systematically to reach a conjecture.

➤ Are you assuming something about the square that is not stated?

➤ Start with a dissection and dissect one of the squares.

➤ Start with a square and build others around it.

If you have done numerous examples, you will have convinced yourself that numbers like

4, 9, 16, . . .

are nice. If you noticed that nowhere does the question say the squares all have to be the same size, then you probably discovered that

4, 7, 10, 13, . . .

are also nice. But are there any others? I KNOW that any dissection contains at least one square, so how can a few others be fitted around it? With any luck you will have discovered that

6, 8, 10, 12, . . .

are also nice, so that putting it all together,

1, 4, 6, 7, 8, 9, 10 . . .

are all nice.

Put another way, it seems reasonable to conjecture that all numbers other than 2, 3 and 5 are nice. But can you convince yourself that 1587, for example, is nice? Here you are looking for some structure which permits you to assert confidently that all numbers bigger than 5 are nice.

TRY IT NOW

STUCK?

Attack
➤ If K is nice, what bigger numbers are also nice?
➤ Try attacking one of your K squares.
➤ What can you deduce from the fact that 6, 7 and 8 are nice?

Extend
➤ The three-dimensional version uses cubes, and gives 'very nice' numbers. It is conjectured that all numbers bigger than 47 are 'very nice' but little is known at present.

The idea of building up a complicated case from preceding cases was mentioned earlier, and *Square Dissection* succumbs to the same approach. One of the basic building mechanisms for square dissections of squares is to dissect one of the K squares in a K-dissection into 4, yielding a dissection into $K + 3$ squares. Thus if K is nice, it follows that $K + 3$ is nice, and so once I know that 6, 7 and 8 are nice, every other larger number is also known to be nice by continued use of the one idea. The Fibonacci sequence, which arose in *Bee Genealogy*, is similar, in that a given term is built as the sum of the two preceding terms. Very often the particular numbers of a sequence are not familiar, but a building scheme can be found which relates each term to preceding ones in some regular fashion. Faced with the task of finding a general formula, mathematicians observe that this same building scheme idea arises frequently, so they turn their attention to finding general methods of translating building schemes into formulae. By such generalization of method, a body of mathematical theory arises.

Square Dissection completes the collection of examples illustrating structure and its role in justifying conjectures. In Chapter 10, *Polygonal Numbers* and *Flipping Cups* are particularly interesting from the viewpoint of structure.

When has a conjecture been justified?

It is important to realize that most conjectures are false, and in many cases the false ones are the most valuable. Despite any apparent contradiction in that statement, the path to a resolution of significance is usually full of false steps, partial or mistaken insights, and gropings toward articulating your sense of what is true. The previous chapters contain several examples, notably:

- *Palindromes* (Chapter 1): the false conjecture that adding 110 repeatedly to 1001 would give all palindromes;
- *Quick and Toasty* (Chapter 2): induces most people to conjecture incorrectly at first;
- *Painted Tyres* (Chapter 4): invites two very plausible but mutually contradictory conjectures.

In each case, it would have been fatal not to CHECK the conjecture, for the process of checking identifies errors and leads to better conjectures. The trouble is that if a reasonable conjecture is discovered after a long struggle, it feels so obviously right that it is hard **not** to believe it. There is a lot of emotional energy invested. As a result, it is easy to be less than critical in CHECKING. How then can you be sure that a conjecture has been adequately checked and convincingly justified? The short answer is that you can rarely be absolutely sure. The history of mathematics is full of false arguments. But you **can** learn to be critical in useful and positive ways.

As indicated in the last section, justifying has to do with revealing an underlying structure or relationship that links I KNOW with I WANT. Once you think that you have found that link, it is a matter of stating it carefully and clearly. As with conjectures about WHAT, your conjecture about WHY may need several modifications, and I recommend three stages:

> convince yourself
>> convince a friend
>>> convince a sceptic

The first step is to convince yourself. Unfortunately that is all too easy!

The second step is to convince a friend or a colleague. This forces you to articulate and externalize what may seem obvious to you, so that the friend is provided with convincing reasons for why what you say is true. It is often helpful to rehearse the most illuminating examples from your specializing, in order to provide your friend with similar background experience to your own. Examples are, of course, not enough by themselves. They may convince your friend that your statements are plausible, but you must justify every step of your argument. For example, it is not enough to say of *Goldbach's Conjecture* or of *Iterates*:

'Do lots of examples and you will see.'

You have to state the structural links which indicate why your conjecture is valid.

Even if your friend is convinced, it is not enough! The third step is to attempt to convince someone who doubts or questions every statement you make. I like to add force by using the word 'enemy'. Learning to play the role of enemy to yourself is an extremely important skill, if only because other suitable enemies may be hard to find!

To see how an internal enemy might function, consider *Patchwork* from Chapter 1. The investigation led to a conjecture about the WHAT:

Two colours always suffice.

Several attempts were made to find a rule for colouring any diagram with just two colours, and a number of conjectures were made and discarded. They were all trying to get at the WHY. With systematic specializing on small diagrams, a method that SEEMED to work gradually emerged:

> When a new line is added, some of the old regions get cut into two parts. Keep all the regions (new and old) on one side of the new line the same colour as before. Reverse all the colours on the other side.

The WHY is not yet fully answered however, because it transfers from

> *WHY are two colours sufficient?*

to

> *WHY does this method work?*

My explanation to a friend might go like this:

> The whole square is properly coloured because regions adjacent along the new line were intentionally coloured differently, and regions adjacent along old lines were previously coloured differently and so remain coloured differently.

Most friends would be convinced, but the enemy needs more. Reading over the explanation, an internal dialogue might now occur:

Enemy: Why are regions adjacent along the new line coloured differently?

Me: Just before the new line was added, those regions were parts of one single region, so they were the same colour. Now one section lies on each side of the new line so they will be coloured differently.

Enemy: But the colour reversal might have made two of those adjacent regions the same.

Me: No, the new line cuts old regions in two.

Enemy: How do you know that regions adjacent along an old line will be differently coloured? Might not they get mixed up with the new line?

Me: Because they were differently coloured to start with, and they were both unchanged, or both changed.

Enemy: How do you know they were both unchanged or both changed?

Me: Because being adjacent along an old line, they cannot lie on opposite sides of the new line.

Enemy: Why?

Me: In order to be on opposite sides of two lines, two regions could only touch in at most a point. They couldn't be adjacent.

The internal enemy is on the brink of following a never-ending chain of WHYs, which is unhelpful scepticism. It is tempting to see some of the questions as indicating that the enemy is basically stupid or, more charitably, unnecessarily nasty, but in fact the questions are not so stupid if you are genuinely sceptical about the way a plane is divided into regions by straight lines. Indeed, the line of questioning is similar to questions asked about the 'obvious' generalization, namely that any curve drawn in the plane without intersecting itself, and joining up at its ends, must, like a line, divide the plane into two regions. Pursuing these WHYs led mathematicians to a vast array of ideas and a fresh perspective (called topology) which sheds light on many different parts of mathematics.

To be helpful, a question must pinpoint a specific weakness. Nevertheless, it is possible for the questions to go on for a long time, forcing me back to very basic principles about straight lines and regions. A proverbial line must be drawn somewhere but it is impossible to make categorical statements about precisely where. Flaws in generally accepted arguments have often appeared when WHY was asked a little more persistently than usual, thus stimulating new mathematical ideas and perspectives.

Patchwork is a nice example because there is a related question:

> How few colours are needed to colour any map so that any two regions adjacent along an edge (not just at a few points) are coloured differently?

In the nineteenth century it was conjectured that four colours would suffice, and an argument stood for many years until someone tried to generalize it and found a flaw. After nearly a century an argument finally emerged, but it had so many steps that a computer was needed to carry them all out, thus opening up further questions for mathematicians about how to check the validity of such a long argument. After all, the computer program might have contained an error or a false assumption. The enemy has a lot of work to do in this case, but the principle remains the same: question the reasoning, and look for fallacies or unstated assumptions.

It may seem that the injunction

convince an enemy

is rather strange, and perhaps overstated. However, it reflects closely the way new mathematical results are accepted by the research community. Following up an insight, an argument is formulated and tried out, perhaps on paper or verbally to a colleague. Several versions later, when various weaknesses have been probed and patched, a paper may be submitted for publication. This version is read critically by at least one expect in the field (the enemy!). Each version tends to become more abstract and formal, trying to be precise and to avoid hidden assumptions of informal language, but incidentally causing the reader to have to work harder at decoding the original insight and the sense of what is going on. As long as the published remarks convince the mathematical community, they are deemed to constitute a justification. Even so, it sometimes happens that an argument is published and accepted, yet years later a fallacy or unstated assumption is found.

Developing an internal enemy

It is not always easy to find a suitable enemy who will patiently but sceptically look at your work, so it is useful to learn to play that role for yourself. Apart from the benefit of not having to seek out some acquaintance, an internal sceptic of your own can play an important role in various aspects of mathematical thinking. Chapter 7 goes into the broader issues, while this section looks at how to become your own best enemy. There are three useful habits which will develop and strengthen your own internal enemy or sceptic.

1 Get into the habit of treating statements as conjectures. The effect will be to alter your perspective from mathematics as a subject in which everything is right or wrong, to mathematics as a discipline of modifying and checking until a convincing justification has been found.
2 Get into the habit of testing conjectures by trying to defeat them as well as seeking a justification.
3 Get into the habit of looking critically (but positively) at other people's arguments. This will strengthen your appreciation of the need to CHECK, because the cracks in an argument may easily be glossed over, especially if it is your own.

Learning to challenge conjectures and trying to defeat them with nasty examples is not as perverse as it sounds, nor as easy. When conjecturing was introduced in Chapter 4, it came as part of a cyclic process.

Distrusting your conjectures is not just lip-service to fallibility. There may indeed be an error, but it may also be that by trying to pick holes in a conjecture you begin to see **why** it cannot be beaten and so must be true. It is curious that belief and disbelief carry quite different perspectives. Looking for why something is true may yield nothing, when seeking to disprove it may reveal what is going on. An instance of this occurred for me when thinking about *Painted Tyres* (Chapter 4). I could not see WHY there would only be one series of marks. It was not until I intentionally disbelieved it and tried to calculate the distance

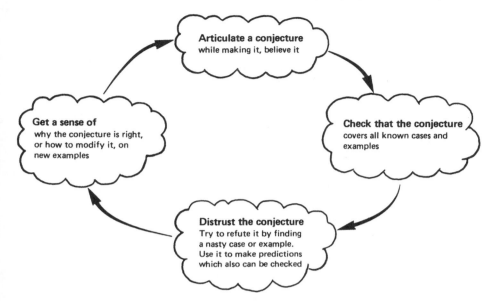

between the front tyre marks and the back tyre marks that I made progress. On extremely recalcitrant problems mathematicians take the age-old advice:

> Believe it true on Monday, Wednesday and Friday.
> Believe it false on Tuesday, Thursday, and Saturday.
> On Sunday take a neutral stance and look for another approach!

One of the best demonstrations of the need for alternating belief and disbelief is based on a Victorian parlour game.

Eureka Sequences

One person writes down a rule which generates three-element sequences of numbers and provides one sample sequence satisfying the rule. The remaining people offer three-element sequences and are given yes/no responses according to whether they do or do not satisfy the rule. All offerings are displayed. When someone thinks they know the rule they shout EUREKA! They are then permitted to offer further sequences which will help everyone discover the rule.

Warning: This game only works if the rule-maker chooses extremely simple rules!

TRY IT AS SOON AS POSSIBLE

A favourite example of Peter Wason (Wason and Johnson-Laird, 1972), who put the game into this format, is to provide 2, 4, 6 as a sample. This suggests some rules, all of which accept the same sequence, but which are **not** his rule. For example:

> 3 consecutive even numbers;
>
> 3 even numbers in increasing order;
>
> 3 numbers summing to 12;
>
> 3 numbers in increasing order at least two (one) of which must be even.

The only way to test these conjectured rules is to try to **refute** them by offering sequences which the conjectured rule would reject to see if the actual rule also rejects them. Most people only offer confirming instances of their rule and so never even have the possibility of discovering that their rule is incomplete. Thus, having conjectured

> 3 even numbers

it is important to offer sequences that are **not** three consecutive even numbers as well as some that are. The rule that Peter Wason uses is **not** one of the ones listed, but any three consecutive even numbers will be accepted by it. To be successful at this game it is essential to recognize and state clearly your conjecture, then systematically test all of the details – in this case evenness and increasing order at least. With this strategy a refutation can enable a better conjecture to replace it.

One of the significant features of *Eureka Sequences* is that unless the rulemaker confirms your conjectured rule, you will never know for certain that your conjecture is correct. The reason is that there is no knowable structure in I WANT (the hidden rule), and no context for the rule you are trying to discover. For example, the rule

> 3 even numbers apart from the triple (22222, 44444, 66666)

is unlikely to be discovered by anyone! All the evidence would point to the rule

> 3 even numbers

and it would be a compelling conjecture, but nevertheless false. This makes the game analogous to scientific investigation, in which the 'laws of nature' always remain conjectures.

A good way to sharpen your critical faculties is to look at other people's arguments, and try to decide whether you are convinced. *Iterates* (page 78) provides one such opportunity.

One person observed that all that was needed (I WANT) was to show that starting with a number N, ultimately the process will yield a number smaller than N (since this would force the process to hit 1 eventually).

The reasoning began by noting that all numbers are either even, or one more than a multiple of 4 ($4M + 1$) or one less than a multiple of 4 ($4M - 1$). The reasoning then proceeded as follows:

If N is even, the next number is immediately smaller.

If N is of the form $4M + 1$, it gives $6M + 2$ (which is even), then $3M + 1$, which is smaller than $4M + 1 = N$.

Otherwise N is of the form $4M - 1$, which does not lead very far.

STUCK!

A second person noticed that it is useful to factorize M into a power of 2 and an odd factor, P. Then $4M - 1$ is equal to $P \times 2^T - 1$ and T is at least 2. From this the next iterate is found as follows:

$$3N = 3(P \times 2^T - 1) = 3P \times 2^T - 3 \text{ (note that } 3P \text{ is odd)}$$
$$3N + 1 = 3P \times 2^T - 2$$
$$(3N + 1)/2 = (3P \times 2^T - 2)/2 = 3P \times 2^{T-1} + 1, \text{ which is the next iterate.}$$

Because $3P$ is an odd number, the next iterate is of the form $4M - 1$ provided $(T - 1) > 1$. Otherwise it is of the form $4M + 1$. In other words, after $T - 1$ iterations, a number of the form $4M + 1$ will be produced from the original number.

WHAT DO YOU THINK? (AND WHY)

STUCK?

➤ Have you checked the two sub-arguments?
➤ Have you phrased what they each show in your own words?
➤ How do the two results fit together? Do they really?

For the people involved at the time, the flush of apparent success could easily have overridden the internal enemy. All insights must be CHECKED!

A resolution

The first result shows that numbers of the form $4M + 1$ subsequently become smaller, but they do not necessarily remain in the form $4M + 1$. The second argument shows that numbers of the form $4M + 3$ eventually become numbers of the form $4M + 1$, but in the process they may get larger and larger, so it is not clear that every number ultimately gets smaller **and** is of the form $4M + 1$. This does, however, suggest further lines of attack, for what is needed now (I WANT) is to show that every number of the form $4M + 1$ ultimately gets to a smaller number of this form. Unfortunately this seems to be just as hard as the original question, indeed it is the nub of the question. Nevertheless, out of faltering attempts like these, new sub-goals, new questions emerge. Perhaps, being more specific than the original, one of them will lead to a resolution. Questions sometimes give rise to such a long chain of subsidiary and modified questions that the original question goes out of focus and is forgotten!

Once developed, the internal enemy can be extremely useful during other phases of thinking apart from justifying, because hidden assumptions can block progress at Entry as well as Attack. In the *Eureka Sequences*, looking for hidden assumptions means noticing that everyone begins by assuming that the rule involves even numbers, and questioning whether evenness is really relevant. In *Square Dissection*, it means noticing that nowhere does it say the squares all have to be the same size. Once seen, such hidden assumptions are all too transparent, but while they are hidden they are very well covered! Here are some traditional examples in which hidden assumptions usually impede or derail the thinker.

Hidden Assumptions

1 Nine dots in a 3 by 3 square array are to be joined by four consecutive straight lines, without removing the pencil from the paper.
2 Three men desperate to cross a river encounter two small boys on a home-made raft. The raft will carry only one man or both boys. Can the men cross the river?
3 With six matches, make four equilateral triangles.
4 How few matches are needed to make six squares?
5 How few equilateral triangles are needed to make an annulus (a ring with a hole in it)? (Triangles must be glued together edge to edge.)

These puzzles can be very annoying, and when finally a solution is found or revealed, there can be a strong sense of **Oh No!**, of feeling cheated. The strength of this reaction is a good measure of how strongly the hidden assumption was held.

Notice the distinction between hidden assumption and conjecture. Conjectures are explicit guesses which may be right or wrong. Hidden assumptions are implicit restrictions which may block progress, precisely because their influence is not recognized.

Summary

Often it is easy to conjecture WHAT, but not so easy to see WHY. To answer why satisfactorily means to provide a justification for all the statements which will convince the most critical reader. To achieve this usually requires a strong sense of some underlying structure, the link between KNOW and WANT. A justification is an articulation of the link. Checking the justification to see if it is convincing can be extremely difficult. Cultivating a healthy, positive scepticism of your own conjectures, actively searching for examples which refute the conjecture, and learning to be critical of both your own and other people's arguments are essential. The three levels of convincing:

> convince yourself
>> convince a friend
>>> convince an enemy

will be developed in Chapter 7 in the notion of an internal monitor who encompasses the internal enemy.

The following chart shows how this chapter fits with the earlier ones:

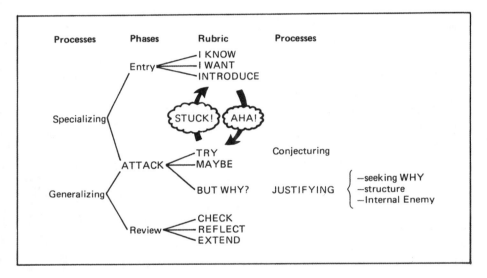

To obtain practice in convincing and justifying, I suggest you return to one or more of the earlier questions where you have been able to conjecture WHAT, but have not yet been able to answer WHY. Try it again, this time searching for underlying structure in the form of links between KNOW and WANT. You might like to try some of the following questions in Chapter 10:

Kathy's Coins	*Liouville*
Flipping Cups	*Polygonal Numbers*
Milkcrate	*Quad-cut Triangles*

See Chapter 11 for other curriculum-related questions.

Reference

Wason, P.C. and Johnson-Laird, P.N. (1972) *Psychology of Reasoning*. London: Batsford.

6
Still STUCK?

This chapter is about preparing for real thinking, the kind that takes place when all the advice offered in earlier chapters has been tried. Your question has now turned into a problem. If you do keep going, you can expect to experience long periods when nothing seems to be happening. Despite appearances, your thinking may still be going on, but below the level of awareness. In order for it to be useful thinking, there are things for you to do!

What happens if you have carried out all the suggestions from previous chapters and still you are stuck? After all the obvious calculations have been done, all the imaginable specific examples have been looked at as systematically as possible, and the work has been checked over for errors, what happens next?

Now is the chance to engage in real thinking. Your question has truly become a problem! Instead of you working on a question, the problem can work away inside you. Before describing in more detail what this means, it is necessary first of all to be sure that the advice of earlier chapters really has been followed. For example, sometimes people shy away from getting down and doing calculations, even when all other avenues seem closed. If you are working on a problem which is difficult, there is no avoiding the work necessary in carrying out extensive specializing, not mindlessly, of course, but always looking for some pattern.

If specializing has not been shirked, then at any time you may find yourself poised at a moment of decision, for you can

- abandon the problem altogether;
- put it aside for a while; or
- keep going.

The decision is not to be taken lightly, if indeed you really have a choice. Very often what happens is that attention is drawn elsewhere and memory of the problem fades or, conversely, the problem may not let go of you! If you elect to abandon the problem, then that is that. It may, however, lie dormant, ready to be triggered back into consciousness by a new fact or a similar question in a different context. If however there is a force inside which does not permit letting go of it then further action is necessary in order to avoid endless repetition of fruitless ideas.

Your activity must now become more subtle, taking the form of active waiting. There are three sorts of activity that are appropriate now:

- distilling the problem to a sharp question;
- intentionally mulling; and
- more extreme specializing and generalizing.

Distilling and mulling

One of the characteristics of the active or demanding problem is that it has been distilled to a succinct essence which you can hold in your mind. The role of all the specializing has been to establish a strong sense of what the **real** problem is and, until it can be encompassed in your mind, it is not ready for mulling. It is difficult to give a concrete example of this state, for it depends very much on past experience and general mathematical competence. It may be that one of the earlier questions, for example *Consecutive Sums* (Chapter 4), brought you to a state in which, after formulating and reformulating the question and permitting it to change its form, you arrived at something such as

> *What property connects all the numbers which can be a consecutive sum in only one way, in two ways, . . .?*

or

> *What property do powers of 2 have that stops them from being a sum of consecutive positive numbers?*

In case you did not have that experience, here is a classical example that might require mulling.

Cut-away

Two diagonally opposite corners of a chessboard are removed. Can you cover the remaining board with dominoes each of which covers two squares?

TRY IT NOW

STUCK?

➤ Have you really tussled with it? Tried smaller boards?

At some point, you will conclude that it is not possible, but perhaps you cannot see why. On rereading the question, the irrelevance of the chessboard strikes you. The real question now becomes, what has the chessboard to do with dominoes? That is what is meant by distillation!

Cut-away is an excellent example of a question which is entirely opaque until you see or are told the idea, and then it all seems so obvious. In fact, it is not worth spending too much time on, so further suggestions are:

> *What colours are the deleted squares?*
> *What does a domino do to colours?*

If you had not seen it before and had not tumbled to the use of colours, then there is likely to be a strong sense of Oh No!, or perhaps Of Course! It seems like a trick, yet when that same insight comes to you without assistance, it comes as an uplifting AHA! At the moment I am concerned with preparing for insight, not with tricks, but it is worth noting that the colouring idea is much more than a mere trick because it can be generalized in a variety of ways to solve a number of quite distinct types of problems (remember *Furniture* in Chapter 4?). Once the colouring idea is found, and noted as a key idea during Reviewing, it is available as a resource when tackling other questions. Suddenly you will detect a similarity to *Cut-away* or *Furniture*, and immediately the colouring idea will surface. If you would like another example, this might be a good time to work on *Kathy's Coins* in Chapter 10.

There are a number of useful activities for distilling a problem, all variations on the same theme. The aim is to articulate as clearly and succinctly as possible the essence of the problem, preferably to someone else. I have often found that the act of trying to express my problem to someone else produces a sudden insight into what is blocking me, without my listener having to utter a single word! It does require an effort on my part to contact the problem and really try to make it clear to my friend. In the absence of a friend, or perhaps in preparation for such advice-seeking, it helps to write down all that you know about the original question, with all the specializing, conjectures and counter-examples. Writing it down makes it easier to check for errors or omissions (far more frequent even among experts than many would care to admit!) and it has a further benefit. If it becomes necessary or desirable to shelve the question for a while, then it can be picked up again at a later date. Indeed, the frequent use of RUBRIC writing should mean that there is no difficulty later in following your train of thought.

The writing down required at this stage of distilling is an interesting combination of the two types introduced in previous chapters: the running commentary produced by RUBRIC writing, and the polished report produced in Review. As in a RUBRIC record, unsuccessful paths need to be recorded so that work that has already been done will not be repeated later. On the other hand, each strand of thought that has been followed ought to be presented in as polished and digested a form as possible, expressed logically and clearly, with each assumption identified. Remember, you are doing this for yourself now, not for someone else. The problem is demanding it!

Having written out what you know, perhaps having sought out a friend and, most importantly, having distilled the essential question, then mulling begins. Despite no signs of progress the prospect is not actually bleak, because by now the problem has become a real friend, like a tune that keeps coming forward in your mind at idle moments, perhaps even taking precedence over other regular fantasies! Attack now turns into a waiting game, waiting for a fresh idea or insight. It is not a passive process, but neither is it active in the direct sense. Rather it requires a fine balance between doing and not doing. Seeking insight is both frustrating and uplifting. If a truly new idea is needed, then dogged persistence at old ideas will not be enough, and may even get in the way by blocking other possibilities. Of course, it is rarely possible to be certain that an old idea will not work, so the temptation to go over old ideas is always present. If you feel desperate to **do** something, then I recommend fresh air and exercise. It at least stimulates the flow of blood and oxygen if nothing else. A change of activity is often helpful because it loosens your grip on the details.

Insight generally comes as a result of the juxtaposition of the question with some new experience, when the two unexpectedly resonate together. Thus the 'doing' which is required takes the form of mulling over the question, rolling it about in the mind like the summer's first strawberry in your mouth. Juggle the components of the question so that new combinations and connections are formed. The 'not doing' takes the form of avoiding working over old ground and letting ideas work together in their own way. There can be at this stage a very definite sense of participating rather than of being the principal agent, as the question ebbs and flows in your attention.

As might be expected, this aspect of thinking has been discussed by many authors for it is in many ways the most intriguing. A number of useful observations have been made which can help to unstick the stuckness, and they all boil down to specializing and generalizing again, but in more extreme forms than before.

Specializing and generalizing

Every so often, despite patiently trying to mull, there is a yearning to do something, to try something different. There are two very reasonable responses. One of them is to specialize in more extreme ways, altering the question by making it more and more specific, or even changing some of the conditions until some progress is possible.

It is not always easy to see how to do this, but the basic message is the lifeblood of most questioning activity:

if you can't solve your current problem, alter it until you can.

Unfortunately some researchers forget the original question in the rush to make progress! None of the questions so far have required very extreme forms of specializing, though the suggestion in *Cut-away* to look at smaller chessboards is an act of desperation that not many people try. If you do try a 2 × 2 board with missing corners, and then a 3 × 3, the basic idea will probably emerge!

Another form of modifying the question to make it tractable is to seek analogies with other questions. This involves a combination of specializing and generalizing because what you do is focus on some particular features of your question. Then you ask yourself if you have ever encountered those features before. If you have been Reviewing your thinking, you will find that you accumulate a rich store of useful and accessible experience, making analogy-seeking more and more fruitful. Put another way, success, with Review of why it was successful, does breed success. Like most of the advice in this chapter, it is very difficult to be precise about analogy-seeking, except by reference to experience.

Some examples from previous chapters include:

- *Fifteen*: the analogy being structural, between triples of numbers adding to 15, and magic squares;
- *Quick and Toasty*: the analogy with *Paper Strip* being that using physical models really can help;
- *Bee Genealogy*: the analogy with *Leapfrogs* being that it is essential to ask WHY your conjecture seems to work, and not be satisfied with a formula that seems to give correct answers.

Many more examples will arise as you tackle questions in Chapter 10, and as you practise Reviewing your thinking on other questions that you encounter.

It is easy to overlook generalizing as a useful activity when specializing has failed to produce an idea. Sometimes problems become clearer when stated more abstractly with the inessential details removed. Generalizing suggests looking carefully at a question and trying to see what role the various constraints are playing. It may easily be that removing one or more constraints will make it easier. *Consecutive Sums* demonstrated this when, having worked out some special cases of two consecutive terms, one or two examples failed to fit what was otherwise a clear pattern. Thus:

$$5 = -1 + 0 + 1 + 2 + 3$$

expresses 5 as the sum of five consecutive numbers, and

$$7 = -2 + -1 + 0 + 1 + 2 + 3 + 4$$

expresses 7 as the sum of seven consecutive numbers. By temporarily not worrying about the positiveness constraint, a resolution emerged.

Hidden assumptions

Having examined all the obvious constraints stated explicitly in the question, there are still likely to be a number of unnecessary but hidden constraints of your own making. These, of course, are the hardest to find, and yet often they are the most significant contributors to being STUCK!

The mind sets like jelly and, becoming set in its way, finds it very difficult to upset that set. Some classic examples of this were given in Chapter 5 under *Hidden Assumptions*. For example:

Nine Dots

Nine dots in a square 3 by 3 array are to be joined by four consecutive straight line segments, without removing pencil from paper or retracing any part of the path.

TRY IT NOW

STUCK?

➤ Have you assumed something unnecessarily?
➤ Is there any restriction on how long the lines can be?

This question has circulated widely, but in Adams (1974) the full extent of the hidden assumptions is revealed. Someone managed to do it in three lines by upsetting one assumption, and someone else did it in one line by upsetting yet another! Think about it! I shall not give any suggestions, as they would only spoil your mulling.

Hidden assumptions are what makes puzzles like *Nine Dots* annoying, but they are not confined to puzzles. They are the basis of our perception. Every so often a hidden assumption is revealed, and this alters the flow and direction of mathematical enquiry. Consider the example:

True or False

For each of the numbered statements in the list, decide whether it is true or false:

1 Statement 2 in this list is true.
2 Statement 1 in this list is false.
3 Statement 3 in this list is false.
4 There are two misstakes in statement 4 in this list.

TRY THEM NOW

STUCK?

➤ Be systematic: if a statement is true, what can you deduce?
➤ Statement 4 needs special care!

The first three statements seem genuinely paradoxical because they can be neither true nor false. The basic idea is that they refer to themselves and such sentences have intrigued people for at least 2000 years. Many people have attempted to resolve them, the usual approach being to try to make the paradox go away by outlawing the statements in some fashion. Since they consist of words, to which meaning is supplied by the reader, it is not entirely certain that their reference to 'this list', and to each other or to themselves is actually legitimate. However, an alternative strategy is to observe that, by trying to resolve them, a hidden assumption is being made: that they need resolving. Kurt Godel questioned this assumption in the 1940s and revolutionized mathematical thinking. Very briefly, Godel constructed a statement about numbers which is self-referent, and which can be interpreted to mean

'this statement cannot be proved'

One consequence of Godel's idea is that some mathematical questions can never be resolved without making new assumptions. Hidden assumptions are always with us! For more details, see Hofstadter (1979).

It is all too easy to become set in our ways, with a variety of unnecessary assumptions. Upsetting is a curious business, precisely because being set in our ways makes it hard to change. Direct assault by carefully analysing each part of a question can only reveal assumptions connected with phrasing. It cannot reveal the possibility of a different perspective. That is why mulling consists of both doing and not doing. A changed perspective or an insight is much more likely to come about after a sincere and direct attack has been abandoned. It comes not from confrontation, but by resonance with another idea after the way has been cleared. It is encouraged by a strongly developed internal sceptic who questions everything.

Here is a question which may provide experience of some of the necessarily rather general remarks made in this chapter, depending, of course, on your mathematical background. I am not going to provide a resolution, but don't give up after only a few hours. Keep going!

Multi-facets

Picture to yourself a length of rope, lying on a table in front of you. The cross section of the rope is a regular N sided polygon. Slide the ends of the rope towards you so that it almost forms a circle.

Now mentally grasp the ends of the rope in your hands. You are going to glue the ends of the rope together but before you do, twist your right wrist so that the polygonal end rotates through one nth of a full revolution. Repeat the twisting a total of T times, so that your mental wrist has rotated through T nths of a full revolution. NOW glue the ends together, so that the polygonal ends match with edges glued to edges.

When the mental glue has dried, start painting one facet (flat surface) of the rope and keep going until you find yourself painting over an already painted part. Begin again on another facet not yet painted, and use another colour.

How many colours do you need?

TRY IT NOW

STUCK?

Entry

➤ **Do not** give up just because you cannot picture it. Find a way!

➤ Specialize. Heavily! Look for physical objects that might help.

➤ Introduce a means to record your simple examples.

Attack

➤ If the three dimensions seem to be making life awkward, try to find a simpler way to represent the essence.

➤ Seek the essence of how to count the facets. Try to simplify methods.

➤ Diagrams!

➤ Have you seen anything similar (in essence) before?

➤ Make a conjecture and test it. Try to combine special conjectures for special cases into one statement.

Summary

This chapter has of necessity been more descriptive than previous ones. Having performed all the routine 'thinking', and felt an inner commitment to keeping going, the best advice is to distil the question down to a form which can be held in the mind and mulled over. Writing out what is known and explaining it to a

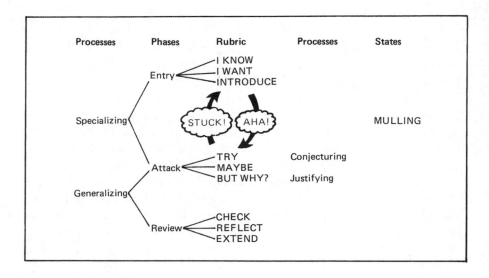

friend often releases a blockage. Otherwise it is a matter of being open to like-nesses with other situations and questioning whether hidden or unnecessary assumptions are causing you to be set in your ways.

To experience the real thinking that this chapter has discussed, I suggest that you return to one of the questions that you previously attempted but did not completely resolve, especially if there is one that nags you. *Circle and Spots* (Chapter 4) nagged me for years. Perhaps that is why it is one of my favourites. You might also like to try generalizing *Nine Dots*.

In Chapter 10 you will find that some of the questions are quite challenging. For instance

> *Folding Polygons*   *Recipes*
> *Moon Struck*   *Wool Winding*
> *Paper Knot*

See Chapter 11 for other curriculum-related questions.

References

Adams, J. (1974) *Conceptual Blockbusting*. San Francisco: Freeman.
Hofstadter, D. (1979) *Godel, Escher, Bach: An Eternal Golden Braid*. London: Harvester.

7

Developing an internal monitor

Each of the previous six chapters has offered advice on how to think mathematically, but having advice printed in a book is not very useful when you find yourself stuck in the middle of an investigation. Trying to select from the book which bit of advice is most appropriate is tantamount to resolving the original question! You may have noticed that whenever possible my suggestions have been in the form of questions or exhortations. The reason is that specific 'hints' remove the opportunity for you to do the thinking, and mask the important points which are the impulses which produced those 'hints'. Furthermore, the whole nature of 'hints' suggests a view of mathematics as a bag of tricks that have to be discovered or revealed. I find this attitude inappropriate and unacceptable. When you are stuck, and even when you are not actually stuck, what you need is a tutor to ask you a helpful question which gets you going again. This chapter describes the process whereby your own internal tutor can grow in strength and effectiveness.

The main thrust of the book so far has been to try to relate my advice to your experience. The purpose of the many questions embedded in the text has been to provide specific, direct experience of thinking so that my advice becomes associated with your experience. In that sense, REFLECTING is probably the most important activity to carry out. It is sometimes said that

   the only way to learn is from experience

but experience alone is not enough. The experience must leave its mark. REFLECTING on the key ideas and key moments intensifies the critical moments of an investigation and helps to integrate their resolution into your thinking repertoire.

Chapter 5 introduced the idea of an internal enemy or sceptic who seeks holes in your arguments. This chapter develops the notion of an internal sceptic into an internal monitor who is always present, and who has a number of roles to play. After describing these roles, the chapter concentrates on the mechanism by which such a monitor can be encouraged to grow.

Roles of a monitor

In previous chapters I have been trying to help you become aware of various aspects of thinking happening inside you, by describing those processes as explicitly as I can. None of the processes or activities I have mentioned are unusual or new. They happen spontaneously inside everyone to varying degrees, often below the level of awareness. By becoming aware of them, and seeing how effective they can be in appropriate circumstances, they should begin to happen more frequently and more intensely than before. Now I am going to offer a different perspective. Thinking activities will be described as if there is an independent agent inside you, monitoring what you are doing. This monitor acts very like a personal tutor who watches you and asks pertinent questions, with the advantage of being privy to your thoughts as well as your actions. What sorts of thing can this monitor do?

1 Keep an eye on calculations to make sure that they remain relevant to the question. If long tedious computations draw you away from the actual question into a side street or blind alley, a gradually growing reluctance to continue indicates that the monitor is stirring.

2 Keep an eye on the execution of a plan (TRY. . .) to make sure that it does not drift too far off course. A growing reluctance or unease signifies that the monitor is beginning to act.

3 Recognize generalizations, however tentative and imprecise, as **conjectures**, and distinguish I KNOW from I WANT and from MAYBE.

4 Evaluate ideas as they come along to see if they are worth pursuing. Rejecting lots and lots of computation in a question that 'seems' not to need it is an indication of the monitor at work. Holding back and considering a plan or idea before jumping in is important monitoring activity.

5 Notice when a state of STUCK has arisen and bring that awareness to the surface thus enabling a change of activity.

6 Suggest reverting to Entry in order to clarify KNOW and WANT, to specialize more systematically, or to introduce an alternative perspective, diagram or notation.

7 Suggest altering a plan of Attack by trying to generalize in a different direction, to seek alternative underlying patterns.

8 Critically examine arguments to see if there are gaps, hidden assumptions or logical errors.

9 Prompt you to Review a resolution, before finishing work.

There is plenty here for the monitor to do and, if that is not enough, there is a further important role which will be taken up in Chapter 8:

10 Look outward and pose new questions stimulated by whatever is going on, whether in mathematical thinking or in ordinary activity.

Notice that I am drawing a distinction between

> being involved in thinking

and

> monitoring the involvement.

This distinguishes the thinking process from awareness of that process and underlines the importance of reflection in contributing to that awareness.

Conscientious reflecting on your own resolutions to some of the questions posed in this book should result in growing acknowledgement of this distinction. At first, as you look back, a few moments will stand out sharply, without needing to be reconstructed. If you persevere you will find that every so often you become aware **in the moment** of your thinking processes.

Emotional snapshots

To encourage the growth of your own internal monitor requires extensive practice with effective reflection. Practice in working on questions is only useful if it builds a reservoir of success. Successfully overcoming being stuck engenders a positive attitude and self image, and it is from the reservoir of success that good ideas and positive attitudes will come in the future. It is essential both to become stuck and to overcome it, because sooner or later you will be stuck and it is then that you most need advice. The source of that advice is effective reflection on previous occasions.

To be able to reflect requires an ability to

> perceive
> > recognize
> > > articulate
> > > > and assimilate

what actually happened, without judgement and without embroidery. Judgement sets up negativity and blocks access to what actually happened, locking you instead into self-justification from which little can be learned. Embroidery and excessive description are what characterize introspection as opposed to moments of inner awareness, and they are highly unreliable. Both judgement and embroidery are caused by the Ego trying to cover up or take over. It is a matter of being honest with yourself, since only you will ever know!

In previous chapters I have stressed the importance of noticing and recording key ideas and key moments. These are events which stand out sharply, with no need to reconstruct them by means of 'I must have . . .' or 'probably I . . .' It is precisely because they make an impact on your feelings that they are useful for developing your monitor. When you recall (but not reconstruct) a key moment, you have access to the feelings involved in that moment. If by reflection you have linked those feelings with actions, that is with thinking processes, recalling the feelings will recall the actions, and with them the specific advice which helped you in the past. If, when working on another question, your feelings resonate with feelings in the past, you have access to helpful advice. Thus there is no need to memorize an extensive list of helpful questions or advice. Rather, it is all locked into your memory in a more or less accessible form, depending on the effectiveness of your reflection.

RUBRIC writing is a device for assisting all this to happen. It highlights events as they take place and provides space for your developing monitor to intervene. If you are constantly jumping into feverish plans of attack then there is little chance for your monitor to be heard! Furthermore, RUBRIC writing provides pointers for stimulating reflection later.

The mechanism for stimulating growth of your internal monitor is conscientious use of RUBRIC writing, and reflection on key ideas and key moments. Because memory of such moments tends to decay with time, it is necessary to learn to 'photograph' the feelings of the moment, so that they can be re-entered. Since the feelings are what give access to the key moments and associated advice, I call such 'photographs' emotional snapshots.

What sort of key moments are worth photographing, and how can it be done? I suggest that you learn to recognize and photograph certain distinctive states which occur while thinking takes place. Each state has a characteristic flavour and, in order to help identify it, I shall offer key words for each one. As the snapshots accumulate, the key words develop a richness which cannot easily be articulated and, in time, the key word becomes a trigger. As soon as the state it describes is recognized, the manifold associations come flooding back to you, complete with relevant advice from similar situations. It is as if someone intimate was helping you – your own internal monitor!

The key words which I use are

> Getting started
>> Getting involved
>>> Mulling
>>>> Keeping going
>>>> Insight
>>>>> Being sceptical
>>>>> Contemplating

To give flesh to those bare bones, I will relate them to the questions posed in earlier chapters. Bear in mind though that it is very difficult to give a precise definition to psychological states. Images and metaphors can be used to convey a flavour, and direct reference to my resolutions may also help. Note that questions you find very easy or impossibly difficult are unlikely to reveal many distinct states because it all happens too quickly, or not at all! I hope that some, at least, of the questions posed in earlier chapters have been nicely in between. Remember that it is the feelings you are out to capture, not the ideas!

Getting started

This state may seem too obvious to be worth mentioning, but there is rather more to it than first appears. In order to get started, it is necessary to recognize and accept that there is a question. Too often the emotions jump in and block recognition or acceptance because of a cultural myth that mathematics is hard and that mathematical thinking can only be done by 'clever people'. I hope that the questions tackled in previous chapters have given you some confidence to counteract this prevalent but erroneous view. GET-TING STARTED is the time for getting a feel for the whole question, for sorting out what the question is really asking, and for becoming familiar with the details.

Initially the question is on the page, outside you. Doubtless some of the questions in previous chapters did not attract you, at least at first. Perhaps you dislike *Matches* type questions (Chapter 5); perhaps you prefer more practical ones. Quite probably the very open-ended nature of *Envelopes* (Chapter 2) and *Fifteen* (Chapter 4) meant that you never got started. Perhaps *Creepy Crawlies* or *Ladies Luncheon* in Chapter 2 seemed too narrow or artificial. Each person at a given moment has a spectrum of interest and only questions falling inside that spectrum will get them started while other questions remain cold and distant. Of course, current interest may be determined by outside forces such as pressure to get a good grade or wishing to get it over with. Whatever the source, it is clear that without getting started, little progress is possible. Think back to earlier chapters to see if you can recall particularly attractive questions and particularly unattractive ones.

DO SO NOW

No matter how justified you believe your opinion is about any particular question, your taste in questions reflects something about you which is worth finding out. What is common to the questions you found attractive? What about the unattractive ones?

GETTING STARTED can be compared to the action of taking a match out of a match box and striking it in preparation for lighting a fire. It is an automatic response to the impulse to light the fire and entails as yet no commitment, no irrevocable act. So, too, a question arises and, like the unlit fire, either resonates with your interest to get started or does not. If the questions always come from an external source and the impetus to get started is external pressure, then there is a good chance that a negative attitude will develop, which will include a heavy resistance to getting involved. Rather, the wish will be to treat the question at arm's length, without commitment. This is the state of many children in schools who want only to get finished as quickly and painlessly as possible. Questions are barely read, and certainly not entered, so there is no chance for a spark of interest. It is no wonder that difficulties arise. After the match is struck it flares, but it may then die out, just as a question may prove not to be attractive. On the other hand it may catch and burn steadily in which case you have STARTED.

The questions and settings used in this book were chosen to try to get around reluctance to get started. For example, questions which contain a surprise often get people going. However, the most interesting questions are the ones you pose yourself. Learning to recognize questions in daily activities, as well as in mathematical contexts, is the best way to broaden appreciation of any question from whatever source. Chapter 8 looks at this in more detail.

Getting involved

The transition from GETTING STARTED to GETTING INVOLVED seems subtle initially, but the two states are really quite distinct. GETTING INVOLVED means activity: head down, hands dirty. The aim is to come thoroughly to grips with the question, to sort out meanings and relationships, to specialize in various ways so that the question comes off the page and gets inside you. In fact, the question becomes your own. Enough work is done on it to distil it to its essence which can be kept in the back of your mind. Technical words are re-expressed in comfortable language with accompanying examples; I KNOW and I WANT are clarified.

In Chapter 1, *Warehouse* usually gets people involved when they discover from one example that in fact the order of calculation of discount and tax does not matter. Having previously treated the question coolly, it suddenly becomes alive and they want to know if it always works and why. So GETTING INVOLVED is not directly associated with initial specializing or other activity, but rather with an increased kind of intensity. This state can be likened to applying a match to kindling, for it captures an involvement between thinker and question (person and fire) not present earlier when everything was treated at arm's length, and it indicates a commitment not present while GETTING STARTED.

Palindromes, or indeed any of the questions which you worked on, will have included a period of time when all attention was on the question and you were increasingly involved in thinking. It would be worthwhile pausing and recalling some of those moments, because from them a great deal can be learned.

Some people very easily become involved, perhaps too involved. The mind races, ideas flow, and then after a while all subsides. If the question succumbs to such a barrage, all well and good, but a great deal of energy can be needlessly spent and good ideas can be lost. One way to counteract this tendency is to use RUBRIC writing, for it slows you down just a bit, records at least some of the ideas (TRY . . .) and reminds you what you were trying to do. With time you can learn to recognize such frantic moments and pause long enough to allow your monitor to survey the scene more soberly. By learning to hold back, you guard against being sucked in by any sudden flare of the kindling, the sudden influx of energy and enthusiasm.

Other people, of a more cautious disposition, like to take their time. For them, GETTING INVOLVED means working their way systematically into the question by careful specializing, drawing of diagrams, introducing notation, re-arranging expressions and so on. Their involvement is just as intense, but more like a steam roller than a racing car. Nevertheless the kindling is being lit and, by learning to recognize the state, they can become more sensitive to associations with other questions that arise inside them and less dependent on plodding.

Apart from the influence of personal psychological type on how people GET INVOLVED, there are interesting observations to be made about the transition from the GETTING STARTED state of recognizing that there is a question, to GETTING INVOLVED. Take for example *Quick and Toasty* in Chapter 2. On first reading, it seemed to me in a vague sort of way that there might be a question here, but it did not appeal to me. Then my attention was focused on it by the influence of a colleague. After a few jottings a surprise was encountered, and I was involved. The kindling had flared, but it took someone else to light the match. The flare was strong enough to catch, and I was involved. Questions which involve practical activity like *Paper Strip*, *Leapfrogs*, *Furniture* or even *Iterates* can be very helpful in getting people involved once any initial reluctance to play has been overcome. Very open-ended questions and situations are the most frightening at first, perhaps because no immediate surprise is seen and the fire may not be well laid. With experience, particularly of extending questions that you have worked on, a wider range of questions will become attractive.

Mulling

If the question succumbs quickly then mulling may have simply been instantaneous or not needed at all. However, there comes a time when a significant question needs to be pondered over. The question seems clear enough and is well inside you, but some new idea or plan is required. This state was described at length in Chapter 6. Its characteristic is a distancing from the question. Whereas the Entry phase means getting more and more involved so that you become one with the problem, the MULLING state of the Attack phase is the reverse. Here you look around for other questions from the past with some similarity or an analogous structure. You try to change the question by specializing or generalizing in new ways to get something tractable. You try to find a new way of looking at the question, representing it by a different diagram or reorganizing your information. A good example of this appeared in *Consecutive Sums* (Chapter 4) in the work leading up to Conjecture 5. My resolution does not convey the amount of pondering on the role of an odd factor which led me back to specialize in a new way. It is typical of MULLING to become distanced from immediate involvement, like hikers pausing on a hill to get bearings before plunging on again. The plunging into a different form of systematic specializing, looking at sums of two, then three, consecutive numbers brought about a new involvement, illustrating that these states are not relentlessly sequential but, rather, often fleeting and recurring.

Another question in which mulling might have arisen was *Fractious* (Chapter 2). Certainly the people in the question got involved immediately and were then surprised to discover they had got it wrong. If you followed the same route, then for a moment you found yourself distanced from the question, seeking an alternative approach. Then a new idea appears. Your monitor's task is to evaluate the new idea before activity begins and you GET INVOLVED again.

So far I have dealt with the distancing aspect of MULLING, yet the connotation of mulling is more to do with soaking or stewing, which describes nicely the act of looking around for another idea. Chapter 6 described in some detail the particular intensity required to soak yourself in a problem, letting it stew in the back of your mind. It is a time when components of the problem are being combined and recombined in various ways. Your task is only to keep the heat at simmer by bringing the question into focus every so often. Having reached this state with a question, you will find that other questions, ideas, techniques, which previously you had not noticed, will suddenly emerge as candidates for helping with this one.

Keeping going

With a challenging problem, the question of whether progress is likely or even possible will arise. Such a moment, and there may be several, marks the possibility of a new relationship with the problem. As mentioned in Chapter 6, it may be appropriate to abandon the problem permanently or temporarily. Often it is sensible to shelve a problem, intending to return to it at some future time. Sometimes, however, the problem will not let go of you and you become aware of a deeper commitment. There is a sense that it can be resolved, or at least that more can be done. There is no rush, no hot intensity as with the flush of GETTING INVOLVED. Rather, there is a sense of harmony or connection with the problem. It has become a friend. Few of the questions posed in this book so far are likely to have engendered a sharp, clear moment of KEEPING GOING, if only because I have always provided indications of a way to proceed and so a moment of KEEPING GOING, of considering giving up and getting assistance will probably have turned into 'let us see what the book suggests . . .'. If you do recognize the KEEPING GOING state, then you have experienced the companionship that mathematical problems can provide.

Apart from the examples in Chapter 6 which were centred on fixation, I had moments of KEEPING GOING in several others. For example, in *Furniture*, I was fully involved moving a square around, and then after a while I seemed to have done all that I could. I had a strong sense that the armchair could not be turned around as required, but I had no idea of a direction to pursue. I was aware of myself surfacing from the conjecture to face not having a plan and the question arose, 'Shall I KEEP GOING?' I kept at it because something deep down said that there must be something going on, and I wanted to find it. Casting around for an idea (MULLING), my monitor said

> 'Write down what you know. Where can one corner of the chair reach?'

Pursuing that (GETTING INVOLVED again) led me to a pattern which suggested what was going on.

Another form in which KEEPING GOING arises is during the carrying out of a plan (TRY . . .) which seems to be getting messy. For example, in *Consecutive Sums*, the plan of representing each of the numbers

1, 2, 3, . . .

in turn as the sum of consecutive numbers culminated in a conjecture, but shed no light on why. The moment when my monitor asked

'Is this working?'

was a local form of KEEPING GOING, by which I mean that it was concerned with a particular plan, not with the problem as a whole. It might easily have grown to encompass the question of whether I should KEEP GOING on the problem, especially if the other idea of looking at sums of two, then three, consecutive numbers had not arisen. This sort of KEEPING GOING arises most frequently when you have embarked on a sequence of calculations which seem to be getting messier than the original question deserves. A sense of unjustified complication arises and your monitor calls a halt for reviewing the state of play.

Your monitor will learn to judge whether things really are messier than necessary or whether it is laziness that is at the source of your reluctance to carry out a lot of calculations. The people who tend to get easily INVOLVED in a headlong rush are also the people who tend to back away from long calculations. They may fail to KEEP GOING just at a crucial point and, instead, cast around for an easier way. By contrast, the people who are more careful and quiet tend to carry on with a plan without the aid of monitoring, carrying unnecessary calculations too far. Such people, when tackling *Furniture*, for example, often spend too long generating data and not enough time looking for pattern or structure in the data.

The decision to KEEP GOING is not easy to make happen. Rather, it comes about as a result of a growing relationship with the question, so that you notice that things have changed rather than decide to change them.

Insight

Very often a resolution appears unexpectedly. After a few calculations or perhaps years of mulling, a pattern emerges which links KNOW and WANT. I choose to distinguish between sudden flares as a twig catches fire (why not try . . . !), and a moment of INSIGHT in which the whole question or a significant part of it seems to fall into place. At such times, it enhances and prolongs the insight to write down AHA! The raising of spirits, closely akin to humour, is a welcome relief after the frustration of MULLING, so hang on to it!

I hope that you have had several examples of INSIGHT as you worked through the questions. *Painted Tyres* (Chapter 4) usually brings an **of course!** when it is seen suddenly and clearly that front and back tyres make the same marks no matter how far apart they are on the bicycle. The moment when the supposed link between the wheels vanishes is a kind of release typical of

INSIGHT. In *Patchwork* (Chapter 1), seeing that it can be done in two colours, even though the idea of building up line by line with colour reversals remains pre-articulate, sometimes comes out as an AHA!

Certainly INSIGHT is a state that comes to you; it is not something you can intentionally bring about. You can prepare for it, however, by GETTING INVOLVED, doing the spadework of specializing and generalizing, seeking similar questions and so on. A curious mixture is required, KEEPING GOING and letting go alternately, almost simultaneously, until something new enters the arena. Many scientists and philosophers have written about this alternation between struggling and relaxing:

> The (philosopher's) stone can only be found when the search lies heavily on the searcher. Thou seekest hard and findest it not. Seek not and thou wilst find.
>
> *from an old Alchemist's Rosarium*

> Chance only favours intervention for minds which are prepared for discoveries by patient study and persevering efforts. *Louis Pasteur*

> Saturate yourself through and through with your subject . . . and wait.
>
> *Lloyd Morgan*

> One sometimes finds what one is not looking for.
>
> *Alexander Fleming*

A moment's insight is sometimes worth a life's experience.

Oliver Wendell Holmes

We cannot carry on inspiration and make it consecutive. One day there is no electricity in the air, and the next the world bristles with sparks like a cat's back.

Ralph Waldo Emerson

Being sceptical

INSIGHTS are frequently incorrect, in part or in full. What I thought I saw clearly may easily have been a mirage, so BEING SCEPTICAL is absolutely essential. There are several different aspects. At the superficial level, an INSIGHT might consist of seeing a pattern, such as powers of two in *Circle and Spots*, which turns out to be defeated by specializing further. A particular calculation may suddenly be suggested, as in *Chessboard Squares*, where a systematic method of counting quickly produces the resolution. More commonly, the insight is tangible but not easy to articulate, and several attempts are needed to say accurately what was perceived. In *Consecutive Sums* it took five attempts to write down as Conjecture 5 what came to me in a flash of insight. All of this is part of BEING SCEPTICAL. Chapter 5 dealt at length with the business of justifying which requires your monitor to BE SCEPTICAL, for no matter how sure you are that you have resolved the question, there is many a slip read 'twixt the cup and lip! Each step of the argument needs careful checking, because it is very easy to be convinced by your own ideas. To do this requires energy, because it is very tempting to stop and think that you have finished. If the excitement of INSIGHT is converted into confidence that the question has been resolved, without BEING SCEPTICAL and checking, then there is a tremendous let-down if you later discover that the INSIGHT was only partial or deficient and that the question remains unresolved. The energy must come from your sceptical monitor, who creates a tension that must be resolved: 'Are you sure?'. The satisfaction and confidence from having reached a resolution, and having checked it, is much greater and more sustained than the uplifted spirits of insight. Although the uplifted spirits accompanying INSIGHT are momentarily exciting, the satisfaction and confidence stemming from a convincing resolution is longer and more lasting.

In a difficult problem it is quite likely that the original question is modified several times in attempts to find a weakness, a way in. BEING SCEPTICAL about your resolution is needed to make sure just what subsidiary question has been answered, and whether further work might be required to answer the original question.

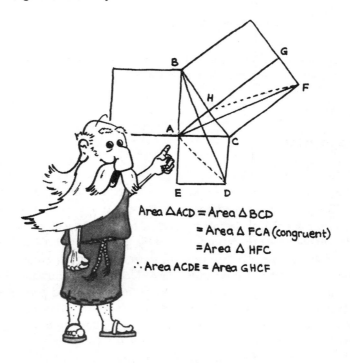

Area △ACD = Area △BCD

= Area △ FCA (congruent)

= Area △ HFC

∴ Area ACDE = Area GHCF

Contemplating

This seventh state is the one I have been emphasizing most heavily throughout the book. It is a calm state of recalling significant events in reaching the resolution, of reading over your resolution looking for the overall picture, and trying to see or set your work in a broader context. It is a supreme form of generalizing in that the current resolution, together with similar questions from your past, are acting as special instances while you seek what is common. Certain states may stand out sharply. A particular mathematical skill may have been useful, so ask yourself what it is about the question which makes that skill useful. You may also be led to ask a general mathematical question leading to a new theory, or a new skill. It is this state, encompassing reflection and extension which stimulates the growth of your monitor.

Summary

Accumulating key moments or emotional snapshots is a long-term process. It also calls for some care. It is tempting to look back over my work and castigate myself for being stupid, for wasting time, or for getting stuck in a characteristic fashion in one of the states. Such remarks made lightly and in humour are one thing, but self-judgements are of negative value only. Our personal propensities are not easy to change. Such change will only come about by carefully and calmly observing them and not by heavy self-criticism. With practice, more and more emotional snapshots will be taken, awareness of your states will increase, and then, when in a particular moment there is also sufficient awareness, change will take place. The other side of judgement is embroidery, an equally negative activity. Reconstructing what 'must have happened', adding details not available in the snapshots, or trying to be too precise or pedantic about distinguishing states serves only to confuse and detract from the growth of your personal monitor.

It is not a quick matter to assimilate the seven states. They are concerned with feelings, which tend to be slippery beasts that are hard to capture in words. It takes time and practice to develop meaning for them. In doing this, it is tempting to expect to see each state come one after another in every question, but psychological states are not like that. They flit back and forth and trying to tie them down too precisely only detracts from the unfolding of meaning. The following chart relates the states to the processes and phases of previous chapters.

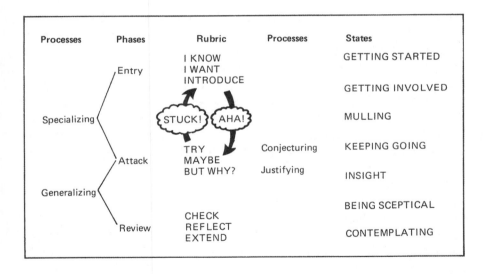

8

On becoming your own questioner

This chapter is about questions. In the earlier chapters I have posed a number of mathematical questions and structured my advice around some thinking questions that you can learn to ask for yourself. Where do all these questions come from, and what use are they? The short answer, which will be elaborated in this and the next chapter, is that mathematical thinking is an attitude, an approach to the world. The mathematical questions I have posed have little value in themselves. They were chosen to elucidate and illuminate the fundamental processes. By tackling them, and absorbing the experiences obtained, you will at the very least have accumulated a rich storehouse of experience that will stand you in good stead in the future. If that future means working on similar questions posed by other people, then, as long as it provides pleasure, the book has been a success. However, that is only the beginning. Much more is possible. Once your confidence begins to grow, you will find that you are more willing to look at problems that are less obviously mathematical, or less highly structured. The reason is that the advice provided in the earlier chapters enables you to undertake specific activities in almost any situation, even though you do not understand the initial question! At the very least you can find out enough to begin specializing. Such an approach indicates that you are developing the attitude of THINKING MATHEMATICALLY. Questions lie at the very heart of that attitude.

This chapter is concerned with becoming aware of the myriad of interesting questions that surround each one of us. The first section considers the wide variation in questions from the specific to the very open, the second considers where questions come from and draws particular attention to what and how we notice, and the third section looks at the forces which blunt natural curiosity and suggests ways of overcoming them.

A spectrum of questions

Most of the questions posed in previous chapters were narrowly specific, leaving little doubt about what sort of answer was wanted. For example:

- *Palindromes:* is every four-digit palindrome divisible by 11?
- *Patchwork:* how few colours are needed . . . ?
- *Ladies Luncheon:* match the first names with the surnames.
- *Quick and Toasty:* what is the shortest time in which three slices of bread can be toasted?
- *Threaded Pins:* how many pieces of thread will be needed in general?
- *Leapfrogs:* what is the minimum number of moves?
- *Goldbach's Conjecture:* every even number greater than 2 is the sum of two primes.
- *Consecutive Sums:* exactly which numbers can be written as . . . ?
- *Bee Genealogy:* how many ancestors does a male bee . . . ?
- *Square Dissection:* what numbers are 'nice'?

However, not all of the questions were as narrow as these. Two which stand out in particular are:

- *Painted Tyres:* what did I see?
- *Envelopes:* how could I make myself one?

Being less specific, these sorts of question are open to a wider range of interpretation than the others. It is not exactly clear what is wanted (indeed there may not be any particular answer), and at first it may even be hard to imagine what a reasonable response would look like, at least until you get started and involved. It is this sort of question that I want to bring to your attention particularly. One way to gain confidence is to practise extending any narrowly specified questions that you come across. Very often a particular or special case wants to be made more general. The aim of this chapter is to set you on the road to generating your own questions!

Questions seem to fall across a spectrum which extends from the very narrow to the highly amorphous. Some of the intermediate types are:

- right answer known and wanted;
- an answer known, and process interesting enough to offer question to others;
- answer suspected, good approach known to similar questions; approach suspected;
- seems interesting but no specific approach suspected;
- question vague or open to generalizing;
- no particular question but situation attractive.

Amorphous open-ended questions are generally found to be more difficult to deal with than narrow well-defined ones, despite the greater freedom offered. It seems that the freedom introduces uncertainty with its associated lack of confidence. Furthermore, there is a big difference between noticing a question yourself and being offered a situation in which someone else has identified a question.

When it is clear from the nature and wording of a question that the questioner already knows the answer, then the thinking can easily turn into a form of competition. Can I find a solution as quickly and as elegantly as the questioner? On the other hand, questions which are less definite offer the possibility of pursuing your own directions, based on **your** interests and on what **you** discover. Open-ended investigations often involve considerable time being spent simply exploring, looking for patterns or surprises with no specific goal in mind other than a general interest, or curiosity to find out what is going on. Once a specific question or conjecture formulates itself, there is no pressure to pursue it if a more interesting one appears. This is in contrast to a specifically posed question which tends to direct and constrain your thinking. Furthermore, thinking about imprecise questions is fuelled by your interest whereas thinking about narrow questions is usually stimulated by outside pressures or by competition.

Even when an investigation has been broadly based, it helps to formulate specific questions on which to focus attention. The difference is that now the questions are your own, and not someone else's. Specializing and generalizing then produce sub-goals, which contribute to the original question. Building up layer upon layer of such sub-questions provides experience of recognizing and formulating particular problems, and of seeing how problems cluster together to give greater insight. That is why I have been encouraging you to extend your results to more general contexts. Even the narrow questions I have been posing are capable of extension.

> Only when a result fits into a wider context do you really begin to see its significance.

> Extending questions is a good way to begin noticing and posing your own questions.

Some 'questionable' circumstances

Both *Envelopes* (Chapter 2) and *Painted Tyres* (Chapter 4) occurred in the course of ordinary events, and I posed them to you more or less as they arose inside me. Although we each come across lots of potentially fascinating questions every day, they are usually not noticed, and even more rarely articulated.

In Chapter 7 I suggested that it is surprise or contradiction which provokes thinking. These arise when something changes, and when I become **aware**, however subtly, that things **have** changed. For example, a rearrangement of furniture, a change in car design or haircut causes some people to notice and comment while others remain unaware, at least at the level of articulation. The kind of questioning which I am advocating is a form of noticing.

Here is an example.

Seesaw

At a play-ground with my children I was struck by the fact that the seesaws were all resting horizontally. Now when I was a child, seesaws were very simple devices consisting of a board on a fulcrum. At rest, one end was always high in the air. Immediately I felt an impulse to inspect these new-fangled ones.

Already the questioning had begun. Before actually looking, I thought about what I knew and what I might find, and I was aware of having a conjecture. It was not formulated in words, and later I regretted not stopping myself and making it more definite and precise. Having inspected the mechanism, several questions arose:

> Why that particular configuration? In other words, suppose someone asked me to design one like it, what questions would I need to answer?
> What is the path of one of the seats?
> Why would anyone bother to change the design?

It suddenly occurred to me that the mechanism had exactly the same form as a rocking horse that I had seen at a friend's house, but with different measurements. I had remarked on the mechanism then, but had not pursued it. The similarity between the two was striking, and certainly accounted for a good deal of my interest.

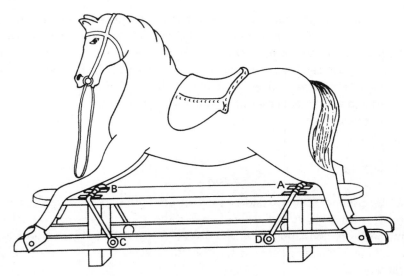

The essential features which made me notice *Seesaws* were the **change** from old to new, and the **surprise** at recognizing the same structure in two apparently different contexts. Of course it helps enormously if you have actually seen and played with one of these seesaws yourself. Everyone responds most readily to surprise and possibility arising from their own experience, and this usually means manipulating or playing with something, which may take the form of physical objects, numbers, diagrams or symbols. The important feature is that they are real for the individual. They must have substance, whether in the physical world or in the world of ideas. They must be able to be confidently manipulated either physically or mentally.

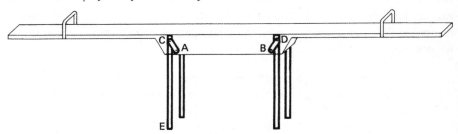

Many engaging questions arise within mathematics which are not immediately practical, but which are none the less intellectually attractive. The root source of their appeal lies in resonances with past experience, and with an intellectual curiosity. Some people seem to prefer practical questions with results that affect their external lives, and other people seem to prefer more abstract questions. However, there is no need whatsoever to be cast into one mould, for there is really very little difference in the process involved. Here is an example of a situation that offers a variety of possibilities to the more abstractly oriented thinker.

Number Spirals

Numbers are written consecutively in a spiral as follows:

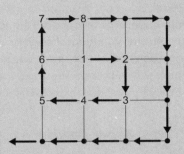

Extend the spiral and write down any questions that occur to you.

TRY IT NOW

STUCK?

➤ Look for patterns and try to make predictions.

Here are a few questions which occurred to me.

> *What numbers appear on the diagonals through 1?*
> *What numbers appear in particular rows and columns?*
> *Where will 87 appear? Generalize!*
> *Where are the evens, odds, multiples of three?*
> *Where are the square numbers?*

The pattern of square numbers struck me particularly, and I wondered why it should be that way. I began trying other spirals based around a few blocked-out squares on squared paper (as here).

	15	16	..	
13	14	1	2	3
12				4
11				5
10	9	8	7	6

		23	24	..		
	21	22	1	2	3	
19	20				4	5
18						6
17						7
16	15	14		10	9	8
		13	12	11		

	9	10	11	12		
	8	1	2	13		
	7		3	14		
	6	5	4	15	16	
		..	21		17	
			20	19	18	

There were many examples in between those three, each one more adventurous than the last. It took a lot of drafting and redrafting to specify what I meant by these more general spirals, each draft being a conjecture as to the most general form of spiral that would preserve the property of the positions of the square numbers. (I am being deliberately vague so as to tempt you to investigate for yourself!) Then I had the task of justifying my conjecture that the same pattern would always occur!

The questions in *Number Spirals* all involve looking for relationships between numbers and positions, with the intention of finding succinct general expressions predicting which numbers will appear in given positions. By contrast, the next situation involves more subtle pattern seeking.

Paper Bands

Take a thin strip of paper, about 11 inches by 1 inch (28 cm by 2.5 cm) and fold it as shown.

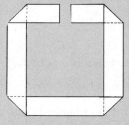

Join the ends to make a band.

DO IT NOW – SEEK SOME QUESTIONS!

Paper Bands arose one day as I was fiddling with some thin paper strips. Various ideas occurred to me, but this one proved particularly interesting **to me**. What I discovered was that the band I formed had a twist in it, and it occurred to me to wonder whether I could predict the number of twists just knowing the sequence of over and under creases.

In most cases mathematical thinking focuses on pattern of some kind, though pattern must be interpreted broadly. A configuration such as a seesaw, a geometrical diagram or a sequence of numbers may resonate with past experience and prompt such questions as

In how many ways?
What is the most/least?
What is the underlying structure?
Will the same technique work more generally?
Why is it like that?

Why is it not like that other situation?
Why does that happen?
What patterns are there here?
Where do these numbers come from?
What happens next?
Can I predict what will happen in general?

More generally, the questions

Of what is this situation a special case?
What is going on here?

indicate the desire to remove inessential detail and concentrate on essence. Such questioning is really the manifestation of an attitude rather than a collection of typical questions, and I shall examine the components of this sort of attitude shortly.

There are no special places to look for questions, indeed quite the contrary. We spend most of our time avoiding them! Often they arise from necessity:

I need to hire a car. Is a daily or weekly rate best for my purposes?
Should I replace my car this year or next?
What is the cheapest way to send Christmas parcels abroad, one big
   package or several small ones?

Sometimes they emerge from more complex personal or social decision making:

Where should I spend my holidays?
Should I change jobs?
Should I extend my mortgage, or pay it off more quickly?

They also arise from intellectual curiosity:

The date 01.11.10 is a palindrome (English system day.month.year). When
   is the next one? What about in the American system (month.day.year)
   or in the International system (year.month.day)?
How many different types of wallpaper patterns are there?
How far apart should street lights be placed?

Once you begin to notice them, there are a great many questions to which mathematical thinking can contribute.

Noticing

I have said that what we notice is change and surprising juxtapositions. Since these depend on our background experience, training, interests, knowledge and current psychological state, each act of noticing is unique to the noticer. For example, an architect and a musician will notice different aspects because of their training. When I return from visiting a new place I often notice references to the place in newspapers and magazines. It has happened too frequently to make it plausible that it is mere coincidence, so I conclude that my new interests are making me aware of things I previously did not notice. Before I repaired a roof I never noticed the variety of shapes and pitches, but now due to my new knowledge I often notice details overlooked by others. A state of confidence and calmness makes it possible to notice and remember details that pass unremarked during exuberance or depression. Because of all these factors, reading other people's noticings gives but a pale reflection of the original energy provoked.

Noticing **can** be improved, strengthened and broadened. It requires only a **wish** to notice, supported by recording **what** is noticed. See Mason (2002) for more details. However, it is not just a matter of walking around asking idle questions. Questions arise as a result of an action inside.

Noticing something like *Seesaw*, *Number Spirals* or *Paper Bands* involves becoming aware of change. That awareness results in an action between the 'before' and the 'after' states. The before and after components are not sufficient, however, because there has to be something which connects them together. Most of the time noticing comes about because some sense-impression, some new perception such as a horizontal seesaw suddenly (but randomly) becomes juxtaposed with a variety of past impressions or perceptions such as a memory of the old-fashioned seesaws. The juxtaposition of impressions is what is experienced as surprise or awareness of change. In other words, the person is aware of an inner tension, conflict or contradiction.

This kind of action depends on new and old impressions somehow becoming juxtaposed. Most of the time the new impression (horizontal seesaw) is simply registered somewhere at random in memory, but nowhere 'near' the old impressions (the old-fashioned seesaws) and so noticing does not take place. Yet we all have within us the necessary data for questions, in the form of mutually contradictory impressions stored in separate 'compartments'. Sometimes a new impression suddenly (but again randomly) forms a bridge between two mutually contradictory memories, and this too produces awareness in the form of tension. It may even reach articulation as a question.

Just because an inner tension produces a question, it does not follow that the question-tension will be resolved. The energy may be released in some fashion such as laughter or activity, thereby avoiding the issue. Even if the tension does emerge positively as a question, the entire action is essentially random.

It is not necessary to depend on random juxtapositions, however. A new impression and some old stored impressions can be brought together intentionally by a wish to understand. The new impression acts directly on the old data, mediated by the intention. Tension is generated and a question emerges. It seems sensible, therefore, to cultivate the intention to notice and to question, to be interested in and to wonder about. Noticing is not so much something that we **do** as it is the result of an attitude or intention.

Obstacles to a questioning attitude

There is a temptation to say

'I am not a questioning sort of person.'

but this response is just one of several ways we all have of avoiding the issue. Anyone can become 'such a person' if they really want to. It is a matter of adopting an active, enquiring attitude to the world. Just as conjecturing is less of an activity and more of an attitude to the ideas that I have or to the statements that others make, so, too, questioning is an attitude, an approach to life.

Another reason often given for not asking questions is:

'It's no use asking questions when I can't answer them.'

Deciding that you cannot answer a question before you even get started is a sign of lack of confidence, but not a valid reason for failing to begin. The whole thrust of this book has been to provide specific concrete activities to carry out whenever you recognize that you are stuck. If you have taken the advice to heart you should find that your confidence in tackling questions is growing. Confidence comes from success, and from knowing WHAT to do, even if you have no idea what is really going on or how it will turn out. Both of these sources of confidence rest on an underlying active approach to the world, which has much in common with the attitudes that generate questioning. Care must be taken to distinguish between an active approach and excessive activity. Some people insulate themselves from asking questions by being extremely active, to the extent of never reflecting. This is excessive activity of a negative kind which does not promote mathematical thinking, even though the activities may be largely mathematical. It is true that such a person may develop mathematical talents to a considerable extent, but I believe it is more important to become aware of the thinking processes than simply accumulate solutions to particular problems.

Furthermore, I have stressed how important it is to learn to accept being stuck, and to try to learn from that experience. There is nothing wrong with being unable to make progress on a question, and there is a tremendous value in tussling with it, rephrasing it, distilling it, mulling it over and modifying it in various ways. By doing this, you make it possible in the future to recognize fresh information or new techniques that will help you make more progress. It will also prepare you to seek expert advice, to ask sensible and probing questions, and to make use of the replies you receive.

One other factor in not asking questions is mental laziness. Like all forms of laziness, of course, it means that the person is on the boundary of possibility. Just as it is the case that when I am feeling a general fatigue with no particular impulse to do anything, as soon as someone comes along and suggests a fresh activity I suddenly find all my energy return, so that, the very act of getting started on a question brings about the possibility of getting involved, a state in which lots of energy is generated and all lassitude vanishes. Thus, not wanting to be bothered to ask questions is like painting over a door that you have never even tried to open. You have no idea what you are missing!

The thrust of my argument is this: those who say that they are not the questioning type have lost contact with the very natural curiosity they had as young children. Not only do we impress upon our young generation that asking questions is not popular, we also convey an atmosphere of tension at not being able to answer questions. Thus we set up the attitude

'It's no good asking questions I can't answer.'

On the contrary, questions that I know I **can** answer are generally less interesting than questions I am unsure about. Questions that I **know** I can make no progress on tend to be classified as uninteresting, but why? An attitude of hopelessness can be generated, for example, by a car that stops working. I can only leave it to an expert. And yet many men and women have found that, after a gentle introduction, car maintenance is not so frightening as it seems. It also has beneficial side effects, since it makes a big difference to have a good idea of what needs to be or is being done to your car. Here is a more detailed example.

A friend's child has a flute which suddenly started playing one note incorrectly. The parents, quite handy mechanically, deduced that one of the springs had become dislodged during cleaning but were unwilling to do anything. The stated reason was that they did not want to harm the flute further. I suspected that there was also some lack of confidence in the child's analysis of the problem. I compared our two flutes physically and aurally, and the child and I worked out which key was not functioning. It took two seconds to alter the spring which was in a different position to all the others, and the fault was corrected. The reason for relating this anecdote is that I had no particular knowledge of a flute's construction apart from a general idea of how mine worked. However, I had the confidence to have a look. Note that I was not confident that I could **fix** it. On the contrary, I was aware that there were things I could **do** (I began by specializing!) which would get me started and involved. At the very least I would expect to be able to say to a flute expert just where the fault lay.

Curiously, it was the same morning that I had with great reluctance tackled our car whose timing had been set and reset without success by a garage after a major service. I decided to look at the only thing I really knew much about, the sparkplugs, and lo and behold, one of the plugs had a gap that was much too narrow. The timing problem was resolved. There is no doubt in my mind that my relief and elation at not having to approach the garage yet again carried over into the rest of the day and helped with the flute. This illustrates to me the cumulative effect of success-confidence.

It is unfortunate that success is usually measured in terms of reaching a goal, with the concomitant that not reaching the goal is seen as failure. That is why I have emphasized that being stuck is honourable, and the remarks above seem to me to indicate that our ever-present search for confidence and support breaks down precisely because of the nature of our goals. Intending or demanding to get **the answer**, and, worse, more elegantly and quicker than anyone else, is asking for failure eventually.

A goal which is framed more in terms of an attitude than an end product seems much more likely to yield confidence without fear of failure. If active participation in thinking is valued most highly then confidence and security can flower. Success in the context of a question can then be phrased in terms of wishing to understand it, change it, clarify it and see it in a wider perspective. It is not necessary to require an answer. Indeed, the most fruitful questions are usually the ones that cannot be answered! This is true not only in philosophy (e.g. What is the meaning of life?) but also in mathematics.

Summary

A questioning attitude can be acquired, or perhaps more accurately can be released from bondage, by expressing and affirming an intention to notice and question. The essential components seem to be

- noticing questions when they arise;
- knowing some things to DO when you get stuck;
- being satisfied with clarifying and perhaps formulating a conjecture;
- genuinely wishing to learn about the world around you and about yourself.

We notice the unexpected, the changing. The new picture on the wall will be more like wallpaper a year from now. The challenge is to see things freshly. Every so often there is a gap in the flow of inner chatter and outer stimulation; then there is room for asking. Even this articulation is misleading, because by the time that I am aware of a question, the 'questioning' has already begun. The action is taking place. The tension is working itself out. My wish is to participate!

The alternative is accepting everything around me as if it were mere background, taking every possible route to avoid uncertainty, and accepting everything without challenging, without asking why and how.

I find that one single question forms the basis of my attitude. At various times

What is going on here?

pops into my head, and sets off a questioning sequence. For me every question is a specialization of this one. It may easily not have the same force for you as it does for me, because each person must identify their own particular generalization!

What is peculiar to mathematical questioning? Typical mathematical questions are of the form:

> *How many . . . ?*
> *In how many ways . . . ?*
> *What is the most/least . . . ?*
> *What properties does . . . have?*
> *What is the same about (several different events, facts, situations)?*
> *Where have I seen something like this before?*
> *What is the essential idea here?*
> *What makes this work?*

Some people enjoy working on problems posed by others, perhaps because they like the security of knowing that it can be done and that somebody else found it interesting. Others like to work on significant well-known unsolved problems. Still others prefer investigating openly, posing their own questions. There are also many whose main enjoyment comes from assimilating the resolutions of others, refining them and making them accessible to a wider audience. All of these roles are valuable. A balanced approach is probably the healthiest!

The following questions in Chapter 10 are either very open, or are capable of being considerably extended.

> *Flipping Cups*
> *Die Rolling*
> *Square Take-away*

See Chapter 11 for other curriculum-related questions, any of which can be considerably extended.

Reference

Mason, J. (2002) *Researching Your Own Practice: The discipline of noticing*. London: Routledge Falmer.

9

Developing mathematical thinking

My aim in this book was to start you on a voyage of rediscovery of your mathematical thinking. I say rediscovery because the same processes are observable in the way in which young children explore and understand the world and, in particular, learn to speak (Gattegno, 1963). I have merely reminded you that you **can** think mathematically, by uncovering, dusting off and bringing the processes to awareness. As your awareness of the processes of thinking grows, so does your range of choices. Now that the journey has begun, you can continue on your own to a deeper understanding of mathematical thinking and the attitudes which encourage it. One of the exciting outcomes of learning to expose, use and develop your own mathematical thinking is that you become sensitive to the mathematical thinking of other people. You can offer to others the tools which you are collecting to help yourself. The methods used in this book to acquire mastery over your own mathematical thinking can be transferred to help develop it in others. These methods are dependent, however, on being able to provoke, support and sustain the approach. This chapter is about how to do that and therefore has particular relevance to teachers, parents and indeed anyone in a position to assist the thinking of others.

I will begin by reviewing and reflecting upon mathematical thinking and how it can be improved. Then, I will consider what is required to provoke, support and sustain it in yourself, and in others.

Three factors influence how effective your mathematical thinking is:

1 your competence in the use of the processes of mathematical enquiry;
2 your confidence in handling emotional and psychological states and turning them to your advantage;
3 your understanding of the content of mathematics and, if necessary, the area to which it is being applied.

This book has concentrated on the first two factors, not because a knowledge of mathematical content is unimportant, but because it usually hogs the stage. Often it is presented as the **only** important factor. Drawing attention to the processes of enquiry and the emotional and psychological states they provoke, and focusing on these factors, seems to me to be a necessary part of helping people towards a more useful and more creative view of mathematical thinking. Furthermore, an over-conscientious concentration on mathematical content can obscure the mathematical thinking that was responsible in the first place for the derivation or application of particular aspects of mathematics. For example, in *Threaded Pins* (Chapter 3) a number of specializations and a series of conjectures led to an articulation relating number of threads with *pins* and *gaps* by means of the greatest common divisor. Greatest common divisor is something that appears on the syllabus of most school mathematics classes so I could easily imagine *Threaded Pins* being presented formally as an exemplification of this piece of content. The opportunity to derive the power of this idea by using mathematical thinking would then be submerged in the abstract application. The evidence of the negotiation of meaning and the recognition of constraining factors which are present in the informal enquiry would be lost in a closed presentation. I chose questions which required a minimum of mathematical background so that, for once, your attention could be focused away from **what** you are learning about particular areas of mathematics and redirected to the processes fundamental to successful mathematical thinking.

Improving mathematical thinking

The plan for improving your thinking has concentrated on these two distinguishable but nevertheless intertwined factors:

- processes of enquiry;
- emotional states.

I began by introducing you to certain processes that underlie mathematical thinking:

> specializing
> > generalizing
> > > conjecturing
> > > > convincing

which were discussed principally in Chapters 1, 4 and 5. Although they may seem obvious as a basis for mathematical thinking, for the novice they are far from automatic. It is not enough simply to tackle questions in order to generate competence in these processes. Despite apparent simplicity, they are subtle notions which become intimate friends only after persistent identification and specific attention to their use and practice. This is as true for improving your own mathematical thinking as for trying to encourage it in others.

More specific still was the advice that was built into the RUBRIC words. A very sensible way to proceed when stuck on a question is to direct your attention by trying to investigate thoughts prompted by such key words as:

- What do I KNOW?
- What do I WANT?
- How can I CHECK?

Chapter 2 in particular was concerned with this. Many detailed suggestions have been given in the text and illustrated in resolutions. They show how to set about answering these and other similar questions, and about implementing the four processes. Although I have taken care to structure much of this advice through the use of a RUBRIC (see, for example, the chart at the end of Chapter 2), I have not suggested that it should be learned in any direct way. The advice is too complex for that and, in any event, mathematical thinking is **personal**. The most reliable source of advice is your own experience, and this can be focused by associating key ideas with emotional states. The important thing for you to establish is your own RUBRIC, the one which works for you. I particularly recommend clarifying KNOW and WANT at every stage. I have tried to emphasize that improving mathematical thinking depends upon:

- tackling questions;
- reflecting on that experience.

This approach of **practice with reflection** is the one that I recommend you to follow as you continue to develop your own thinking or as you assist the growth of thinking in others.

Mathematical thinking is not only improved by learning how to conduct an enquiry, but also by recognizing and harnessing to your advantage the feelings and psychological states that accompany it. At the most basic level, there are negative emotions to control.

In particular, in Chapters 3 and 6, I drew attention to the state of being stuck. I pointed out that it does have positive features and there are ways of capitalizing on these. Recognizing being stuck as an ordinary and acceptable state associated with the thinking process changes the focus from a panic about oneself

HELP! I'm stuck!

to an aspect of investigation

STUCK! What can I do about it?

In Chapters 5, 7 and 8, I suggested that one way to overcome being stuck is to develop an internal monitor that will conscientiously observe, question and challenge. Again, the growth of such a monitor hinges upon practice with reflection. Recognizing your emotional states is necessary, so that the monitor can

interpret and use them

by, for example, recalling actions associated with AHA!, or distancing you from the paralysing effects of being stuck, and

control them

to enable you to take steps that you might otherwise resist, like slowing down, checking or convincing.

Few would dispute the necessity of practice to build up an understanding of what happens in tackling a question, to develop a repertoire of helpful strategies and to become competent in using them. But, without reflection, practice can wash over you, leaving no permanent marks. Most of us have had this experience. I can remember at school my teacher frequently instructing me to draw a diagram, but, in my rush to get an answer, I regarded this advice as merely a teacher's quirk. As a result, experiences were lost to me which could have been useful on future occasions. I now see the process of representing a question as offering me time to engage more closely at the same time as I am gathering information. It is this reflection which is an essential accompaniment to practice. By reflecting on experience and encapsulating the most vivid in an emotional snapshot, action can be stored in a significant and readily accessible form.

The technique of improving your thinking by practice with reflection is a simple one, but it needs time. The rapid question/answer format of many mathematics classrooms is the antithesis of the time and space upon which developing mathematical thinking depends; so is the notion that mathematical thinking is the product of practising on repetitive mathematical examples, each done as quickly as possible. Instead, the practice demands ample time for tackling each question independently and the quality of the reflection depends upon the time to review thoughtfully, to consider alternatives and to follow extensions.

Looking back over this book should support this suggestion that time and space are a required part of developing mathematical thinking. Time taken to review allows a different set of expectations to be generated. The richness of a question can be explored many times from different points of view. Its power to act as an analogy or template for future questions is also thereby increased. For example, in Chapter 4 you met *Painted Tyres* for the first time. It cropped up again in Chapter 5 in a discussion of seeking to refute a conjecture. There it was again in Chapter 7, providing an example of insight. Each subsequent meeting offered an opportunity to think about *Painted Tyres* differently. The same question thus provided many opportunities for redirecting mathematical thinking:

What more can be found out?

is a very different orientation from

Now you have finished, try something different.

Effective reflection depends upon a new attitude being developed; for thinking to grow, the important measure is not the number of questions done but the quality of thought put into tackling the question and into reviewing the effort. Making this shift in attitude in yourself, or more especially in those you influence, is an important but difficult task addressed in the next two sections.

Provoking mathematical thinking

Although I have emphasized the pleasurable aspects of thinking, that does not mean that it is easy. To persist with thinking to the point that you can learn from it requires considerable perseverance, encouragement and a positive attitude to getting stuck.

The use of social pressure or authority often generates the outward appearance of thinking. For example, a respected teacher may suggest an activity and pupils embark upon it. However, unless a change comes about inside the pupils, nothing richer than the mechanical application of procedures and the use of previously learned rules is likely. A gap, a space, an emptiness must be perceived and accepted in such a way as to trigger a start to the thinking process. Surprise, contradiction, an unexplained occurrence can act as such a trigger. In *Warehouse*, the natural expectation that the order of calculation does matter is contradicted by the first specialization. Surprise! Previous indifference or only mild interest turns to curiosity. Attention is focused. Thinking can begin. In *Painted Tyres*, the question itself is possibly unusual or expected enough to create the necessary curiosity. The question seems simple and natural. Usually a conjecture is quickly made and accepted in keeping with the question's simplicity. Discussion with others reveals mutually contradictory conjectures and further thought is provoked.

Both of these examples show instances where experience leading to one view is challenged by new information, or a different impression. Once a contradiction or surprise is externalized, as in the case of conflicting conjectures in *Painted Tyres*, a different sort of action takes place. The new impression acts on the old view and, provided that the thinker is curious and wishes to resolve the conflict, thinking takes place. That is what I mean by saying that thinking is provoked by a gap. The unexpectedness of conflicting data jogs us from ordinary existence into awareness, however fleeting, of a gap, of emptiness. This generates a tension which can be expressed:

- cognitively as 'I don't understand';
- emotionally as a feeling of tautness, excitement or even fear;
- physically in a tightening of muscles.

Do not confuse the tension which is created between a

> QUESTION   and   ME

with the tension often provoked by

> I MUST (to get a good grade, etc.)

and

> I CAN'T (don't know what to do)

QUESTION–ME tension has a sense of excitement which stimulates further interest. MUST–CAN'T tension is often due to lack of confidence, inability to see an immediate answer, or the panic generated by classroom pressure.

Real thinking requires getting involved and mulling over a period of time. This cannot happen when the emphasis is on getting **the** answer quickly and correctly and then turning to something else more pleasurable. The deep and lasting pleasure that comes from understanding does not have time to happen!

The question is entered by focusing on what I KNOW and WANT, thereby beginning to bridge the gap by doing something constructive. As the question is engaged, the tension moves from QUESTION–ME, to KNOW–WANT. The new tension generates sequences of AHA! and STUCK! like sparks jumping across the gap, ceasing, then jumping again in succession. Each AHA! is a sense of connection between KNOW and WANT which may, upon checking, amount to nothing, so back to mulling to await a new spark. During the various mullings, the content of KNOW and WANT may change, with the hope of making the gap narrower. Thus, the original question may undergo changes, it may be specialized in a variety of ways, generalized or altered altogether. My attention may turn to a similar or analogous question which appears to offer more possibilities.

If tension is too great, I may not get started. Sometimes, after working at a question, the gap between KNOW and WANT actually widens, as for example when an apparently simple question like *Iterates* turns out to be difficult. When a gap is too wide for the spark to jump from QUESTION to ME, the tension can disappear and I lose interest. On the other hand, excessive tension can impede thinking or lead to an over-rapid attack headlong up a blind alley. Each person is different. You must learn to recognize and flow with the consequences of your own tension, and to respect the flow of those with whom you work or teach.

Keeping the thinking going through periods of frustration is not easy. It requires recognition of an unresolved conflict, paradox or inconsistency as a personal challenge and confidence to accept that challenge. A teacher who understands this and is sensitive to the interests of students will choose questions which can provoke thinking.

Supporting mathematical thinking

No thinking takes place in a vacuum. The cognitive and emotional atmosphere affect your thinking, whether you are aware of it or not. To be an effective mathematical thinker, you need confidence to try out your ideas and to deal with your emotional states sensibly. The foundation of confidence rests on experiencing the power of your thinking to increase your understanding. Only reflective, personal experience can do this.

Reflecting on your successes, even if they are only partial, builds up confidence. It is especially important for a teacher to recognize how essential confidence is and to create a supportive environment where some success comes to each pupil. Working in groups is helpful and choosing suitable questions is essential.

An atmosphere where confidence can grow is necessary, but not sufficient. To flourish, mathematical thinking requires not only nurture, but also extension. Three components in particular create such an atmosphere. They are necessary to your own mathematical thinking. They are critical if you are in a position to affect the mathematical thinking of others. A MATHINKING ATMOSPHERE is

- questioning;
- challenging;
- reflective.

Since confidence is the key, the necessary attitude can be summed up as **I can**.

QUESTIONING:

I can $\begin{cases} \text{identify questions for investigation} \\ \text{query my assumptions} \\ \text{negotiate meanings of terms} \end{cases}$

CHALLENGING:

I can $\begin{cases} \text{make conjectures} \\ \text{seek justifying or falsifying arguments} \\ \text{check, modify, alter} \end{cases}$

REFLECTING:

I can $\begin{cases} \text{be self-critical} \\ \text{expect and assess different approaches} \\ \text{shift, renegotiate, change direction} \end{cases}$

From their earliest years, children can develop confidence to question, challenge and reflect. But they must be encouraged and reinforced in this. Their curiosity needs nurturing, their investigative potential structuring, their confidence maintaining. If you have not had such experiences, you need to create them for yourself. Cultivate the art of asking the questions which an over-conscientious reliance on methods and facts may have blocked. In Chapter 8, I outlined the different levels of questioning which can help you build the confidence and skills to feed your mathematical thinking. Use these in challenging your own and others' assertions, demanding substantiation and cultivating a healthy scepticism. Remember that for questioning to be valuable, it must be appropriate. If you are in a position to affect the learning of others, note how frequently you create the opportunity for **them** to think, to articulate their own questions, to challenge conjectures and to reflect on what has or has not been established.

Look back over the book and you will see that this atmosphere has been implicit throughout. I have taken QUESTIONING, CHALLENGING and REFLECTING as the natural way to proceed. With this present reminder now fresh in your thoughts, go back and see if a little more questioning, challenging or reflecting would make a difference to some of your resolutions.

For example, in *Palindromes* (Chapter 1), I could have rested with my conjecture that every palindrome can be obtained from 1001 by adding 110 successively. My internal monitor was active and suggested a counter-example. My conjecture was challenged and I reconjectured more successfully. In a classroom of children, such challenging and counter-challenging can be encouraged as part of the discourse of enquiry and the habit out of which an internal monitor grows can be established. It is worthwhile to formulate conjectures as conjectures and leave them in that state, even if you cannot see where to go to test them. Just to have the experience of making conjectures is valuable. There is a significant difference between an atmosphere which expects children to supply correct answers, and one where conjectures are made, challenged and modified, where the demand is

Convince! (First yourself, then ME!)

Where an atmosphere is built around such questions as:

How do I interpret that?
Why do I assume that?
When is that so, and not so?
What do I mean by that?

mathematical thinking is supported.

Sustaining mathematical thinking

Thinking mathematically is not an end in itself. It is a process by which we increase our understanding of the world and extend our choices. Because it is a way of proceeding, it has widespread application, not only to attacking problems which are mathematical or scientific, but more generally. However, sustaining mathematical thinking requires more than just getting answers to questions, no matter how elegant the solution or how difficult the question. The aim of this book has been to demonstrate the particular contribution which mathematical thinking can make to the growth of self-awareness. If you have become involved in the previous chapters, have actively endeavoured to record your process comments by RUBRIC writing and, most importantly, have taken the time to pause and reflect on key ideas and significant moments, you are probably more aware now than before of what happens as you tackle problems. However, awareness means more than that. Awareness is a bridge connecting disparate areas of knowledge, information, experience, perception and feeling to each other and to the world outside. Like the self-referent questions which you met in Chapter 6, awareness operates on itself. I need to be aware of the existence of processes which help me but while I am learning to use those processes I cannot simultaneously learn content. However, once I am aware of both content and processes, separately and interacting, my awareness expands on to another level and I am simultaneously aware of being involved and of the consequent psychological states which my involvement induces.

Increased awareness does not just happen. It has to be fostered, tended and itself built upon in a conscious way. I have chosen mathematics as an appropriate place to concentrate upon awareness. At first sight, to many, especially those who are not comfortable with mathematics, this might seem like an extraordinary, even absurd, choice. Awareness has always seemed to be the particular prerogative of the Arts, rather than the Sciences. However, mathematical thinking has a very special contribution to make to awareness in that if offers a way of structuring, a direction of approach, a reflective power as well as creative and aesthetic potential. Whether the focus of questioning is practical and related to the material world, or more abstract dealing with number, patterns and structure, as in this book, resolving brings a sense of pleasure and confidence, provides space and time for awareness to grow and allows a closer and more effective relationship to develop between the personal and the material world.

Picture mathematical thinking on a helix which loops round and round. Each loop represents an opportunity to extend understanding by encountering an idea, an object, a diagram or a symbol with enough surprise or curiosity to impel exploration of it by MANIPULATING. The level at which manipulation begins must be concrete and confidence inspiring and the results of the manipulation will then be available for interpretation. Tension provoked by the gap which opens between what is expected and what actually happens provides a force to keep the process going and some SENSE of pattern or connectedness releases the tension into achievement, wonder, pleasure, further surprise or curiosity which drives the process on. While the sense of what is happening remains vague, more specialization is required until the force of the sense is expressed in the articulation of a generalization. Articulations do not have to be verbal. They might well be concrete, diagrammatic or symbolic but they will crystallize whatever is the essence underlying the sense which has been achieved as a result of the manipulations. An achieved articulation immediately becomes available for new manipulating, hence the wrap-around of the helix. Each successive loop assumes that the thinker is operating at a deeper level of complexity. The connectedness of the loops always permits the thinker the opportunity to track back to previous levels and therefore to revise articulations that might have begun to wobble.

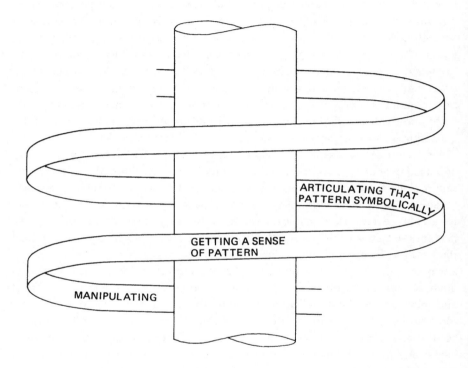

This picture displays how the processes and the emotional states are linked together dynamically. The moment of engaging with a question at Entry demands specializing. Manipulating objects that are concrete (confidence inspiring) for you provokes a gap which stimulates Attack. Conjecturing and convincing together lead to a sense of what underlies the question, and out of that sense grows the articulation of a generalization. An opportunity is then created for Review when a match is made

> back – between the achieved generality, the original state and the experience in Attack;

and

> forward – between the achieved generality and further questions which it provokes.

The impetus is then created for further manipulating at the next level of complexity.

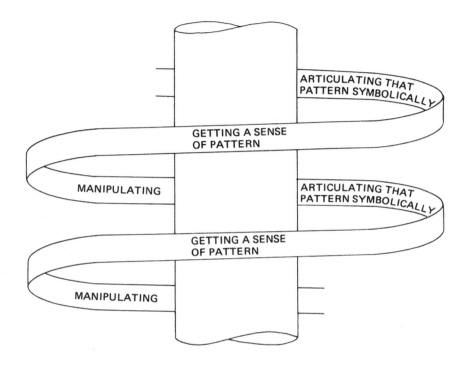

Think back to *Furniture* in Chapter 2. Activity began by manipulating a physical model, a mental image or a diagram. The aim of manipulation was to find out if the chair could be moved as requested. A sense of impossibility emerged, articulated as a conjecture:

'No! I don't think it can be done'

But why? Am I sure? Might not a longer route achieve the result? Manipulating now had a stronger aim, to answer WHY? Perhaps a more abstract notation can replace the physical model as I seek a reason for impossibility. I introduced arrow symbols to show me where the chair reached, but they were not clear and required reference back to the sense I was trying to capture, particularly until I realized that it was horizontal and vertical, and not arrows, that I was after. Then, the symbols became extensions of my thinking so I manipulated them confidently and eventually they formed an articulation of my resolution.

At any point, if faced with a momentary lapse of understanding, confusion or total bafflement, the sensible action is to backtrack down the helix, appealing to a sense of pattern, and to more concrete examples. Specializing, with an eye to discerning meaning at the point of difficulty, is then an attempt to grip the helix firmly and climb back up on more solid foundations. Unfortunately, when symbolic expression is concrete for you and you are attempting to aid the thinking of someone else, it is tempting to jump quickly to a succinct, precise expression and so to use special terms and symbols not realizing that the person you are trying to help may still be struggling further down the helix. Discerning the differences between Manipulating, Getting a sense of, and Articulating, as they operate at different levels of the helix helps to increase sensitivity to where you, or the student you are trying to help, really are. It points to where an abstraction is not based on sufficient experience to be backed up with a sense of meaning. It can also help in assessing where a gap in understanding has arisen and so point to where direct assistance is required. It explains why, whenever we are trying to understand something new, we say 'Give me an example', 'Show me'.

Summary

The view I have presented of THINKING MATHEMATICALLY can be summed up as the answers to a number of questions:

WHAT is mathematical thinking?

● A dynamic process which, by enabling us to increase the complexity of ideas we can handle, expands our understanding.

WHAT do we use to do this?

● Specializing, generalizing, conjecturing and convincing.

HOW does it proceed?

● In phases – Entry, Attack, Review.
● Associated with emotional states – getting started, getting involved, mulling, keeping going, insight, being sceptical, contemplating.

WHICH are the phases to underline?

● Entry – because it lays the foundations for Attack.
● Review – because it is the least acknowledged and most educational.

WHAT improves mathematical thinking?

● Practice with reflection.

WHAT supports mathematical thinking?

● An atmosphere of questioning, challenging, reflecting, with ample space and time.

WHAT provokes mathematical thinking?

● A challenge, a surprise, a contradiction, a perceived gap in understanding.

WHERE does mathematical thinking lead?

● To a deeper understanding of yourself.
● To a more coherent view of what you know.
● To a more effective investigation of what you want to know.
● To a more critical assessment of what you hear and see.

The essence of this chapter is contained in five statements:

1 **You** can think mathematically!
2 Mathematical thinking **can** be improved by practice with reflection.
3 Mathematical thinking is provoked by contradiction, tension and surprise.
4 Mathematical thinking is supported by an atmosphere of questioning, challenging and reflecting.
5 Mathematical thinking helps in understanding yourself and the world.

These were stated at the beginning as the five assumptions on which this work is based.

If you wish to continue your own thinking, Chapters 10 and 11 provide more questions. However, the real test of your success will be when you find yourself thinking mathematically about the problems which you come across day by day. All thinking involves both pain and pleasure: the pain of incomprehension and struggle to understand, and the pleasure of insight and convincing arguments. Mathematical thinking is no exception. I hope that the approach suggested in this book has provided a method, a form of 'relaxation exercises', to help in the control of pain and to provide an impetus which will result in enough pleasure to make the experience worthwhile, and recurrent.

Reference

Gattegno, C. (1963) *For the Teaching of Mathematics*. New York: Educational Explorers Ltd.

10
Something to think about

These questions are offered as fodder for the growth of your mathematical thinking. Resolutions have intentionally not been provided, so that Keeping Going can be fully experienced. They vary widely in sophistication but you might recall the story of the undergraduate in Chapter 3 (p. 52) before allowing that to bother you. Furthermore, although the advice offered is plausible, it does not necessarily indicate the best way to proceed. You must get your monitor active before embarking on any particular plan!

Arithmagons

A secret number is assigned to each vertex of a triangle. On each side of the triangle is written the sum of the secret numbers at its ends. Find a simple rule for revealing the secret numbers.

For example, secret numbers 1, 10 and 17 produce

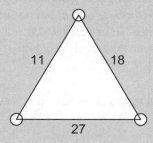

Generalize to other polygons.

Entry

➤ Algebraic symbols will help, but a physical model (say, beans hidden in matchboxes) may help more. Specialize in two ways – with arithmagons where you first choose the secret numbers, and with arithmagons where you give yourself only the side numbers.

Attack

➤ An algebraic result needs thoughtful interpretation to yield a simple rule. What does the algebra really say?
➤ Break up the problem. Some kinds of polygons behave differently to other kinds.
➤ A rule is simple if you could explain it and why it works to a 12-year-old 'enemy'.
➤ Check that your rule deals with impossible numbers on the edges.

Extend

➤ Consider arithmagons on more general arrangements of vertices and edges. For example

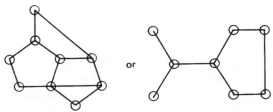

or

➤ Consider operations other than summing.

Black Friday

A Friday the thirteenth is known as a black Friday. What is the most/least number of black Fridays you can get in one year? In a 12-month period?

Entry

➤ Make a conjecture before consulting a calendar.

Attack

➤ Try to find a systematic way to proceed.
➤ What is the least amount of information you need about a particular year in order to work out the number of black Fridays?
➤ Watch out for leap years!

Extend

➤ What about Friday the twelfth, or other numbers? Is there anything special about thirteenths?

(See also *Moon Struck* later on.)

Booklets

Small booklets can be made by folding a single sheet of paper several times and then cutting and stapling. I would like to number the pages before making the folds. Can you tell me how to go about it?

Entry
➤ Try folding a sheet several times, then number the pages without cutting, and unfold it again.
➤ Now specialize systematically.
➤ What do you WANT?

Attack
➤ How are the page numbers on opposite sides of the paper related?
➤ Try and find a general rule which works for any number of folds. How can you communicate it to others most effectively?
➤ What checks can you offer me so that I do not make a mistake?
➤ Specialize to a strip of paper so that all folds are parallel.

Cartesian Chase

This is a game for two players on a rectangular grid with a fixed number of rows and columns. Play begins in the bottom-left-hand square where the first player puts his mark. On his turn a player may put his mark into a square

directly above
or directly to the right of
or diagonally above and to the right of

the last mark made by his opponent. Play continues in this fashion, and the winner is the player who gets their mark in the upper-right-hand corner first. Find a way of winning which your great aunt Maud could understand and use.

Entry
➤ Specialize. Choose small grids and play the game.
➤ You WANT to give instructions for
  (i) whether you go first or second,
  (ii) what move to make in response to any possible move by your opponent.

Attack
➤ Work backwards. Look at how you finish rather than how you start.
➤ Where do you want your opponent to be? How can you force your opponent there?

Extend
➤ What if the rules are changed so that the player who plays in the upper-right-hand corner loses?
➤ Would your method of play enable you to play on a huge rectangle without undue counting?
➤ Three dimensions?

Review

➤ Compare with *Taking Matches* (p. 178).

Clocked

When my son was born, my wife and I agreed that if he woke up before 5 a.m., she would go and feed him, but after 5 a.m. I was to go and bring him into our bed. One night when he woke up, my wife looked at the clock in the dark and said it was my turn. Surprised, since it seemed rather dark, I nevertheless acquiesced. Later it turned out that my wife had looked at the clock upside down, and mistaken 12.30 for 6.00. When will the hands of an upside-down clock show a proper clock time?

Entry

➤ Try it!
➤ Conjecture some plausible times and check.

Attack

➤ Try enough examples to reach a conjecture.
➤ But why?
➤ What changes with time, and what stays the same?

Extend

➤ What if the clock is viewed from other angles?
➤ What if it is viewed in a mirror?

Coin Slide

Select three large coins and three small coins, and place them in a row so that consecutive coins are touching, and so that large and small alternate. A move consists of sliding a pair of adjacent coins to a new position in the row, without interchanging them. Can you, by a sequence of moves, put all the large coins at one end and the small at the other? Oh yes, consecutive coins in the final row must all be touching.

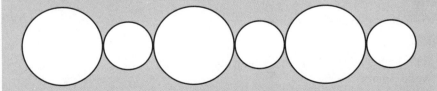

Entry

➤ Specialize to smaller numbers of coins.

➤ Clarify KNOW and WANT. Aim to get a strategy for any number of coins.

➤ Introduce a notation.

Attack

➤ Find some way to represent allowable moves, and possible positions.

➤ If it cannot be done, produce a convincing argument! Remember *Leapfrogs*?

Extend

➤ Try three sizes of coins.

➤ Try moving three consecutive touching coins at a time.

Cube Painting

How many different cubes can be made such that each face has a single line joining the mid-points of a pair of opposite edges? Same question for a diagonal stripe.

Entry

➤ Be clear on what the faces look like (what you KNOW).

➤ Be clear on what 'different' means (what you WANT).

➤ Introduce a picture showing all six faces!

Attack

➤ Have you got all the possibilities?

➤ Have you got repetitions?

➤ Convince an enemy!

Extend

➤ Try a tetrahedron . . .

Cycling Digits

I have in mind a number which, when you remove the units digit and place it at the front, gives the same result as multiplying the original number by 2. Am I telling the truth?

Entry

➤ Be clear on what you KNOW.

Attack

➤ Start somewhere! Make an Assumption!

➤ Keep going!

➤ Write down what you are doing to generate new digits.

➤ A single number is not an adequate resolution.

Extend
➤ What is the smallest number with this property?
➤ Replace 2 by other numbers. Must they have only one digit? Compare patterns.
➤ Move the leading digit to the units place instead.

Desert Crossing

It takes nine days to cross a desert. A man must deliver a message to the other side, where no food is available, and then return. One man can carry enough food to last for 12 days. Food may be buried and collected on the way back. There are two men ready to set out. How quickly can the message be delivered with neither man going short of food?

Entry
➤ Specialize. Try smaller deserts.
➤ Do it physically or diagrammatically.
➤ Introduce a convenient way to record where the men are and what has happened to the food.

Attack
➤ Ten days to cross and return is good going, but not good enough!
➤ Is there any point in some men making more than one trip out?

Extend
➤ Is the nine-day desert the largest that can be crossed without allowing extra time for building up food supplies in the desert?
➤ How large a desert can M men cross in minimum time?
➤ In terms of food-carrying capacity, what is the largest desert across which two men can deliver a message and return without starving?

Diagonals of a Rectangle

On squared paper, draw a rectangle 3 squares by 5 squares, and draw in a diagonal. How many grid squares are touched by the diagonal?

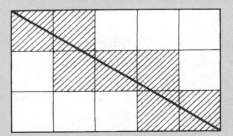

Entry

➤ What is meant by touched? You decide!

➤ Specialize. Be systematic.

Attack

➤ Obviously you can do it by counting, so generalize!

➤ Focus on just the horizontal grid lines. Make a table.

➤ Look for a pattern. Check it!

Extend

➤ What if the diagonal is replaced by a lawnmower cutting a diagonal swatch of grass?

➤ What if the rectangle is divided into rectangles?

➤ What about three dimensions?

➤ What if the grid lines are not equally spaced?

Die Rolling

Obtain a die and some squared paper so that the die just fits on to one square. Place the die with 1 upwards. Mark the square that the die covers with a 1. Now roll the die over one edge and mark the newly covered square by the number showing on top of the die. Experiment.

Entry

➤ What sorts of question arise?

➤ Did you always start on the first square with the die the same way round?

Attack

➤ Can you get any number on to any square?

➤ Can you convince an enemy?

Extend

➤ What is the shortest route from the first square with a 1, to a given square with a given number showing?

➤ Which numbers can you get the right way up?

Divisibility

To check whether a number is divisible by 11, sum the digits in the odd positions counting from the left (the first, third, . . .) and then sum the remaining digits. If the difference between the two sums is divisible by 11, then so is the original number. Otherwise it is not. Why does this work?

Entry
➤ Be sure about what you KNOW. Have you tried simple examples to decode what you are told?

Attack
➤ Try using place value.

Extend
➤ What is special about 11? Can you construct similar tests for other numbers?
➤ What about other bases in place of base 10?

Eggs

Eggs come in a variety of sizes. Which size is the best buy?

Entry
➤ For what purpose?
➤ Get some data!

Attack
➤ State clearly what you want.

Extend
➤ Try it for potatoes, oranges, meat . . .

Fare is Fair

I wish to share 30 identical individual sausages equally amongst 18 people. What is the minimum number of cuts I need to make? What is the minimum number of pieces I need to create?

Entry
➤ Find one way to do it.
➤ Is it the best?

Attack
➤ The particular question is easy to answer. Generalize!
➤ Specialize systematically. Record the results.
➤ Relate the number of cuts to the number of sausages and to the number of people.
➤ Try to find a procedure for calculating the minimum number of cuts/pieces.
➤ It would be nice to have a succinct formula.
➤ Are you sure that you have a minimum? Convince!

Extend
➤ What if the people do not want to share them equally?

Faulty Bricks

Many building bricks are twice as long as they are wide. I need to pile some up until I am ready to use them, and I would like to do it in such a way that each layer is free of any fault line going right across the layer. What sizes can I make the layers?

Entry
➤ Specialize to small layers.
➤ Formulate a more precise WANT.

Attack
➤ Can you build large fault-free layers?
➤ Can you build large layers from small ones? Is there a largest layer which must have a fault?

Extend
➤ In how many ways can a layer be made fault free?
➤ Change the relative dimensions of the bricks.

Finger Multiplication

The following technique was widely used in medieval Europe. Knowing how to multiply two numbers less than 6, you can multiply two numbers between 5 and 10 as follows. Open both palms towards you. To calculate 7×9, say, put $7 - 5 = 2$ fingers down on the left hand and put $9 - 5 = 4$ fingers down in the right. Count the number of down fingers ($4 + 2 = 6$) and multiply together the number of up fingers ($3 \times 1 = 3$) and put the two answers together (63). Does this work, and why?

Entry
➤ Try examples systematically.

Attack
➤ Try to articulate what is going on. It may help to introduce symbols.

Extend
➤ Was it worthwhile learning this method?
➤ Can the method be extended to use toes?

Flipping Cups

My son has opened a package of cups and distributed them on a table, ready for a party. Some are upright and some are upside down. By flipping two cups at a time, can I get them all upright?

Entry
➤ Specialize. Use some cups.

Attack
➤ Are you sure? Can you convince an enemy?
➤ Try small numbers and list all possible positions that you can reach. Look for something common to all of these.

Extend
➤ Consider three or more hands flipping simultaneously.
➤ Consider dials in place of cups, with three or more positions on the dials, perhaps not all the dials the same.

Folding Polygons

Which polygons can be folded (by one straight fold) along a line of symmetry so that the resulting pieces are both similar to the original? What if the fold does not have to be along a line of symmetry?

Entry
➤ Try to find one figure that will satisfy the conditions.

Attack
➤ How many sides can the polygon have?
➤ Work backwards. How can two similar pieces be joined together so that the combination is again similar?

Extend
➤ What if there is more than one fold before demanding similarity?

Fred and Frank

Fred and Frank are two fitness fanatics on a run from A to B. Fred runs half the way and walks the other half. Frank runs for half the time and walks for the other half. They both run and walk at the same speeds. Who finishes first?

Entry
➤ Be clear on what you KNOW. Specialize in order to find out!
➤ Specialize with specific numbers in order to find out what computations are involved.
➤ Diagram?

Attack
➤ Who runs farthest?
➤ Do you really need to know how fast they run?
➤ Interpret everything you know on a diagram.

Extend

➤ Francis joins them and teaches them to jog. Fred now runs one-third of the way, jogs one-third of the way and walks the rest, while Frank jogs for one-third of the time, runs for one-third and walks the rest. Who finishes first? Has Francis helped them to finish sooner or later than previously?

Full-length Mirrors

What is the height of the shortest wall mirror in which you can see both your hair and your shoes at the same time?

Entry

➤ You need to KNOW something abut reflections in mirrors.

Attack

➤ Conjecture: you see more of yourself if you back away.
➤ Try tracing the outline of your face on a steamy mirror.

Extend

➤ What is the least width?
➤ How are the height and width questions related? Why?
➤ Where must you stand to see all of yourself?
➤ What happens if the mirror is not on the wall?

Glaeser's Dominoes

George Glaeser of Strasbourg put a set of dominoes more or less randomly in a flat tray and took a photograph. The exposure was not correct and, although the numbers can be discerned, the positions of the individual dominoes cannot. Can you reconstruct the dominoes?

```
3  6  2  0  0  4  4
6  5  5  1  5  2  3
6  1  1  5  0  6  3
2  2  2  0  0  1  0
2  1  1  4  3  5  5
4  3  6  4  4  2  2
4  5  0  5  3  3  4
1  6  3  0  1  6  6
```

Entry
➤ What do you KNOW about a set of dominoes?
➤ Are there the right number of numbers showing?

Attack
➤ Record your deductions so that you can repeat them and check them!
➤ By systematic. Organize.

Extend
➤ For smaller domino sets, can you pose a Glaeser photograph to give a unique, but not obvious solution?
➤ Can you find another Glaeser photograph that gives a unique solution for a full set?

Gossips

Each evening in a certain village, the old men gather in pairs to exchange gossip about village activities. At each exchange, each one passes on all that he has learned about the day's events. What is the fewest number of exchanges needed so that everyone is up to date on all possible news?

Entry
➤ Specialize to small villages.
➤ State KNOW and WANT clearly.
➤ Introduce a diagrammatic notation.

Attack
➤ Specialize systematically.
➤ Look for general patterns which keep the number of exchanges small.
➤ Convincing a friend that you have found the minimum is quite challenging.

Extend
➤ What if exchanges are replaced by one-way communications?
➤ What if some people will only tell all they know to selected friends?

Half Life

Walking in my home town some years ago now, I suddenly realized that I had been in my job for one-quarter of my life. Perhaps because I was somewhat despondent at the time, I immediately asked myself how long it would be until I had been in my job for one-third of my life.

Entry

➤ Lack of data? Go over carefully what you KNOW.

➤ Do not be rigid about the form of what you WANT!

➤ Would a diagram help? At least try to visualize the question in some way in order to enter it.

➤ Perhaps introduce some letters (as few as possible), but do not be seduced by algebra too soon!

Attack

➤ Write down what you KNOW about what I realized, in convenient notation.

➤ Write down what I WANTED.

➤ Medieval mathematicians would have had no need for symbols!

➤ Express the result in a simple way akin to the original question.

Extend

➤ How long will it be from then until half my life is spent in the job?

➤ If I get fired or retire, how long until one-third (one-quarter) of my life has been spent in the job?

Half Moon

There is a public house in Oxford called the Half Moon, whose sign shows a perfect half moon with a vertical straight edge on a starry night. Something about the sign made me feel uneasy. How do you feel?

Entry

➤ Is it ever possible to see a half moon with a vertical straight line?

➤ When or why not?

➤ Do you KNOW how the moon's phases arise? Are you sure?

Attack

➤ At what angles is it possible to see the straight line at night? In the day time?

➤ Specialize! Get some balloons. Stop the earth spinning.

Extend

➤ Suppose the period of the moon was different.

➤ Suppose the plane of the moon's motion was different.

Handshakes

At a party I attended, some people shook hands. At the end, two people were surprised to find that they had both shaken the same number of hands.

At a party that I gave, attended only by couples, some people shook hands. Upon enquiry, I discovered at the end that everyone else had shaken a different number of hands. How many hands did my spouse shake?

Entry
➤ Work on one at a time!
➤ Specialize in order to find out what you KNOW and what its effect is.
➤ Make some reasonable assumptions about handshakes.

Attack
➤ You need a conjecture to attack!
➤ What one thing does each person at the party KNOW (that is relevant!)?
➤ What are the possibilities for handshaking?

Extend
➤ What would happen on Mars where greetings are always three-person affairs?

Hundred Squares

How few straight lines are required on a page in order to have drawn exactly 100 squares?

Entry
➤ 22 lines assume something about the squares.

Attack
➤ Change the question. How many squares are there in a rectangular grid?
➤ Lots of specializing!
➤ Recall *Chessboard Squares* (Chapter 1).

Extend
➤ Replace squares by rectangles or triangles.

Ins and Outs

Take a strip of paper and fold it in half several times in the same fashion as in *Paper Strip* (page 3). Unfold it and observe that some of the creases are IN and some are OUT. For example, three folds produce the sequence

in   in   out   in   in   out   out

What sequence would arise from 10 folds (if that many were possible)?

Entry
➤ Specialize with paper.
➤ Have you written down a clear definition of IN?

Attack
➤ Be systematic.
➤ Look for a pattern.

➤ What happens to the sequence when you fold the strip an extra time?
➤ What happens to a crease when you fold the strip again?

Extend
➤ What about folding in thirds?
➤ If each crease is unfolded to a right angle and the strip placed on its side, what patterns will you see? Will the strip ever run into itself?

Jacobean Locks

A certain village in Jacobean times had all the valuables locked in a chest in the church. The chest had a number of locks on it, each with its own individual and distinct key. The aim of the village was to ensure that any three people in the village would amongst them have enough keys to open the chest, but no two people would be able to. How many locks are required, and how many keys?

Entry
➤ Specialize.
➤ Try a small village.
➤ Try requiring that any two people can open it but no one person can.

Attack
➤ To help the villagers, you might consider labelling the locks in some way.

Extend
➤ If you count the locks on a particular Jacobean chest, can you deduce the number of people who held keys under such a system?
➤ In a nearby village run on more feudal lines, each villager is rated as to importance (1 is the most important). The squire wishes to arrange that, if a group of villagers wish to open the chest, then there must be at least as many people present as the importance of some member of the group.

Jobs

Three men have two jobs each. The chauffeur offended the musician by laughing at his long hair. The musician and the gardener used to fish with John. The painter bought a quart of gin from the consultant. The chauffeur courted the painter's sister. Jack owed the gardener £5. Joe beat Jack and the painter at quoits. One of them is a hairdresser and no two have the same job. Who does what?

(From: Problem 44.3, *M500 Society Magazine*, 1977, OU Student Journal.)

Entry

➤ State explicitly any assumptions you make.
➤ Introduce a table or other recording device.

Attack

➤ Record your various deductions so that you can CHECK them later.

Extend

➤ Find a way of making up such puzzles.
➤ There should always be a unique solution.

Kathy's Coins

25 coins are arranged in a 5 by 5 array. A fly lands on one and tries to hop on to *every* coin *exactly* once, at each stage moving only to an adjacent coin in the same row or column. Is this possible?

Entry

➤ Specialize to smaller arrays.
➤ Aim to answer for general rectangular arrays.

Attack

➤ What instructions would you give to the fly, assuming that it cannot think ahead?
➤ Can you exploit symmetry?
➤ Can you find a pattern in the good and bad starting points?
➤ How can you convince an enemy that something is possible or not possible?

Extend

➤ What if the arrays have some coins missing?
➤ What if diagonal moves are permitted? Other moves?

Knotted

By how much is a rope shortened when a simple overhand knot is tied in it?

(This also appeared as Problem 6297 in *Mathematical Monthly*, **87**(5), 1980, p. 408.)

Entry

➤ Try various rope sizes.
➤ What are you measuring?
➤ How tight are you pulling it?

➤ You need a definite question!
➤ Formulate KNOW and WANT precisely.
➤ Introduce a schematic diagram.

Attack
➤ Simplify your diagram until you can use it to produce a formula which will estimate the answer.
➤ How well do your predictions fit the data?
➤ Can you improve your diagram or your formula?

Extend
➤ Try other kinds of knots.

Leap Birthday

I was present once when a father truthfully told his seven-year-old daughter that it was his ninth birthday. The daughter asked when they would first have celebrated the same number of birthdays.

Entry
➤ KNOW and WANT! Watch out for hidden assumptions.

Attack
➤ Do not believe your conjecture. CHECK!

Extend
➤ What is the longest period during which father and daughter can have celebrated the same number of birthdays?

Liouville

Take any number and find all of its positive divisors. Find the number of divisors of each of those divisors. Add the resulting numbers and square the answer. Compare it with the sum of the cubes of the numbers of divisors of the original divisors.

Entry
➤ Take it easy, it's not as bad as it sounds!

Attack
➤ Specialize to numbers with few or simple divisors.
➤ If your conjecture is correct for two numbers, is it true for their product?

Match Boxes

Match boxes customarily have their length, width and depth of different lengths. Three such boxes can be assembled into a rectangular block with all three boxes parallel, in three distinct ways. In how many ways can 36 boxes be similarly assembled?

Entry
➤ Be clear on what you KNOW.
➤ Try some literally, or figuratively.

Attack
➤ Specialize to two dimensions.
➤ Specialize to cubes.
➤ Do not be satisfied with your first conjecture.
➤ You need a convincing argument.
➤ Can you build the answer for large numbers from the answers for smaller numbers?
➤ Are there numbers of boxes for which there is only one way? Just three ways?

Extend
➤ Relate the answers for match boxes, cubes, and boxes with two equal dimensions.

Medieval Eggs

A woman on her way to market, when asked how many eggs she had, replied that, taken in groups of 11, 5 would remain over, and taken in groups of 23, 3 would remain over. What is the least number of eggs that she could have had?

  On another occasion she replied that taken in groups of 2, 3, 4, 5, 6 and 7 there would remain over 1, 2, 3, 4, 5 and no eggs respectively.

Entry
➤ Specialize to smaller numbers.
➤ What do you WANT?

Attack
➤ Could any numbers at all have been given in reply?
➤ Be content with a procedure for answering egg questions! A formula may not be possible.

Review

➤ These two questions are typical of puzzles that circulated widely in medieval times.

➤ There are records of similar puzzles which were popular some 2000 years ago!

Milk Cartons

How much cardboard do you need to make a carton to hold 1 litre of milk?

Entry

➤ What shape have you chosen? Why?

Attack

➤ Do not forget the flaps, etc.

Milkcrate

A certain square milkcrate can hold 36 bottles of milk. Can you arrange 14 bottles in the crate so that each row and column has an even number of bottles?

Entry

➤ Depict the crate. Find a way to be able to manipulate bottle substitutes.

➤ Specialize to crates of other sizes.

➤ How many bottles might there be in each row and column?

Attack

➤ Specializing to larger crates might help.

➤ Eventually you will deal with the 36-bottle crate, but what about square crates in general?

➤ Can you build up new arrangements from old ones?

Extend

➤ Try to place other numbers of bottles.

➤ Try rectangular crates.

➤ What is the largest/smallest number of bottles that can be suitably arranged in a given crate?

➤ How many ways are there to place the bottles?

Moon Struck

On Wednesday last a full moon shone on my pillow as I went to bed. When will that next happen?

Entry
➤ Be clear on what you WANT. I wanted it to be a Wednesday at the same time.
➤ Be clear on what you KNOW about the moon.

Attack
➤ Does your conjecture fit with your experience?
➤ Check what you KNOW in an atlas.
➤ What does it mean to say that the moon is in the same position?

Extend
➤ How frequently does Christmas day have a new moon?

More Consecutive Sums

In *Consecutive Sums* (Chapter 4), I asked which positive numbers can be expressed as the sum of consecutive positive numbers. Now I would like to know in how many different ways a given number can be so expressed.

Entry
➤ Be systematic again!
➤ Make use of the insights from *Consecutive Sums*.

Attack
➤ You WANT a pattern linking the structure of a number and the number of representations as a consecutive sum.
➤ Do you KNOW which numbers have a unique representation?
➤ Don't be put off by unnecessary assumptions. Try to integrate the powers of 2 into your thinking.
➤ Try finding all numbers which have exactly two representations as consecutive sums.
➤ Recall *Math Boxes*.

Extend
➤ Try altering the kinds of sum to conditions such as sums of squares or sums of consecutive odd numbers.

More Furniture

Try replacing the square armchair in *Furniture* (Chapter 4) by a sofa which is 2 units by 1 unit.

Entry
➤ Use the strategy that worked in *Furniture*.

Attack
➤ Formulate conjectures, check them and modify them.
➤ Keep going!

Extend
➤ Vary the proportions of the sofa.
➤ Try other kinds of furniture (love-seat?).
➤ Try restricting movements to other angles.
➤ Can you get a general rule to cover all these cases?

Not Cricket

Amongst nine apparently identical cricket balls, one is lighter than the rest which all have the same weight. How quickly can you guarantee to find the light ball using only a makeshift balance?

Entry
➤ Specialize to fewer balls.
➤ What sort of thing do you WANT?

Attack
➤ Do not assume that a particular ball is the light one!
➤ What is the worst that can happen?
➤ Are you convinced that it cannot be done in fewer weighings?

Extend
➤ What if there are more than nine balls?
➤ What if you know only that one ball has a wayward weight?
➤ What if there are two kinds of balls, heavy and light, but unknown numbers of each?
➤ What if the balls are all different weights, and I wish to line them up in order of weight?

Nullarbor Plain

A man lost on the Nullarbor Plain in Australia hears a train whistle due west of him. He cannot see the train but he knows that it runs on a very long, very straight track. His only chance to avoid perishing from thirst is to reach the track before the train has passed. Assuming that he and the train both travel at constant speeds, in which direction should he walk?

Jaworski et al. (1975).

Entry
➤ What does the man KNOW?
➤ If he knew the direction the track ran in, which direction should he walk?
➤ What does he WANT? Be reasonable!

Attack
➤ Conjecture: he walks north (well, why not?)

Odd Divisors

Which numbers have an odd number of divisors?

Entry
➤ What do you WANT?
➤ Write some down and look for a pattern.

Attack
➤ Keep Going until you have a conjecture.
➤ State it clearly.
➤ Now convince an enemy.

Extend
➤ Is there a number with exactly 13 divisors?
➤ Generalize.

One Sum

Take any two numbers that sum to one. Square the larger and add the smaller. Square the smaller and add the larger. Which do you expect will be larger?

Entry
➤ Specialize enough to reach a conjecture.
➤ CHECK all calculations with fractions or decimals!

Attack
➤ You must find a way to convince an enemy.

Extend
➤ Find a corresponding pair of calculations for two numbers that sum to S.
➤ Illustrate One Sum in a diagram of rectangular areas.
➤ Find a corresponding pair of calculations for two numbers whose product is P.

Pancakes

When I make pancakes they all come out different sizes. I pile them up on a plate in the warming oven as they are cooked, but to serve them attractively I would like to arrange them in order with the smallest on top. The only sensible move is to flip over the topmost ones. Can I repeat this sort of move and get them all in order?

Entry

➤ Specialize.

➤ Clarify what you WANT.

➤ Do you WANT to know if it is possible?

➤ Do you WANT to know the least number of flips needed in the worst case?

➤ Perhaps you could be satisfied with an estimate of the least number of flips.

Attack

➤ Specialize systematically, looking for a strategy, and for what the worst case looks like.

➤ Not all questions have neat solutions. Be content with an estimate for the number of flips in the worst possible case.

Paper Knot

Take a narrow strip of paper and tie a simple overhand knot. Gently tighten it until you get a flat, regular pentagon. Why is it a pentagon? Why is it regular?

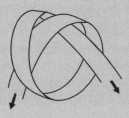

Entry

➤ Try it!

➤ Do it enough times to be sure of what you are doing (what you KNOW).

Attack

➤ Be absolutely clear on KNOW and WANT.

➤ Try to capture the essence on a diagram.

➤ Try following the path of the strip without interlacing it to form a knot.

Extend

➤ Can you make other regular polygons?

➤ Write down explicit instructions for the ones you can do!

Pólya Strikes Out

Write down the numbers 1, 2, 3, . . . in a row. Strike out every third one beginning with the third. Write down the cumulative sums of what remains. Thus

```
1,  2,  3,  4,   5,  6,   7, . . .
1,  2,      4,   5,       7, . . .
1,  3,      7, 12,       19, . . .
```

Now strike out every other one starting with the second, and write down the cumulative sums of the remaining numbers. Recognize the sequence?

This phenomenon was discovered by Moessner (1952). See Conway and Guy (1996) for an extreme generalization!

Entry

➤ Extend the initial sequence.

Attack

➤ Make a conjecture. Convince yourself that it works.

➤ Why does it work?

➤ Try specializing. Begin again with the original sequence, but just do the last set of striking out and see what happens.

➤ Work backwards from WANT to KNOW.

Extend

➤ Place it in a more general context. Generalize the number of striking out sets.

➤ Try starting with other sequences, or varying the strike out rule.

Polygonal Numbers

A number which can be represented as the number of dots in a triangular array is called triangular.

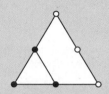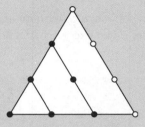

A number which can be represented as the number of dots in a pentagonal array is called a pentagonal number.

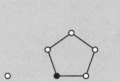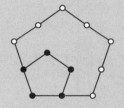

Which numbers are triangular, which pentagonal and, more generally, which are *P*-polygonal?

These were studied by Pythagoras, by Nicomachus of Gerasa (late first century CE) and by every generation since.

Entry
➤ You WANT a formula for the *K*th triangular number.
➤ Specialize to square numbers first. They might be easier!

Attack
➤ Find out how the triangular numbers grow.
➤ Try sticking triangular numbers together.
➤ Try decomposing square numbers into triangular numbers.
➤ Try decomposing other polygonal numbers.

Extend
➤ Which numbers are both square and triangular?
➤ Try counting other arrays of dots.

Quad-cut Triangles

This triangle has been cut into four quadrilaterals and a triangle. Can you cut it into quadrilaterals only? (No new vertices are permitted on the edge of the original triangle.)

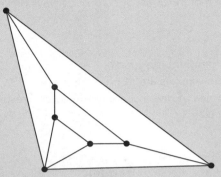

Entry
➤ Specialize by reducing the number of quadrilaterals.
➤ Articulate a conjecture!

Attack
➤ TRY supposing that you had done it. What can you deduce?

Extend
➤ What if other polygons are to be dissected?
➤ What if other polygons are permitted in the dissection?

Recipes

3 litres of orange concentrate were mixed with 5 litres of water to make a drink. Later, 2 litres of orange were mixed with 3 litres of water. Which mix is more concentrated? Consider the following strategy. To compare

   3 Orange and 5 Water with 2 Orange and 3 Water

remove the second from the first and compare

   1 Orange and 2 Water with 2 Orange and 3 Water.

Remove the first from the second and compare

   1 Orange and 2 Water with 1 Orange and 1 Water.

Now you can see that the second was the more concentrated.

Will this strategy always work?

Our source was Alan Bell but see Noelting (1980) and Streefland (1991).

Entry
➤ What is going on? Be clear about KNOW and WANT.
➤ Specialize to other examples.

Attack
➤ What is the essence of the strategy? Does it always preserve what it claims to preserve?

Extend
➤ What if another ingredient is added?
➤ What if three or more mixes are to be compared?
➤ What if several copies of one mix are removed and result in negative quantities in the other mix?

Repaint

The squares of a chessboard are repainted randomly in black and white. Must there be a rectangle all of whose corner squares are the same colour?

Entry
➤ Try smaller chessboards.
➤ Try to show that there need not be such a rectangle.

Attack
➤ Change the question to find the largest chessboard for which such a rectangle need not exist.
➤ Must the chessboard be square?
➤ What do you mean by largest rectangular chessboard?
➤ Specialize by looking at what happens if all the squares in one row are the same colour.
➤ Find a systematic way to specialize.
➤ Does the order of rows and columns matter?
➤ Be clear, when you specialize to particular cases, about what you KNOW.
➤ What might an acceptable answer look like?

Extend
➤ Have you tried three colours?
➤ What about three dimensions?
➤ Can you ever guarantee the existence of a square whose corners are all the same colour?

Reversals

Take a three-digit number, reverse its digits and subtract the smaller from the larger. Reverse the digits of the result and add. Thus

123 becomes 321 and 321 − 123 = 198

198 becomes 891 and 198 + 891 = 1089

What happens? Why?

Entry
- Use a calculator to specialize.
- Make a conjecture.

Attack
- Introduce some pictures or symbols.

Extend
- Have you tried it with four- and five-digit numbers?
- Have you tried it in different bases?
- Change the rule to simply reversing digits and subtracting the larger from the smaller, and repeating over and over.

Right Angles

Given the number of sides of a polygon, what is the maximum number of right angles it can have?

Fielker (1981).

Entry
- Clarify KNOW and WANT.
- Must the polygon stay in the plane?
- What do you mean by right angle? Inside or outside?
- What kinds of polygon do you admit? Self-crossing?

Attack
- Specialize.
- Look for principles that help you build large polygons with many right angles.
- Make conjectures. Do not believe them. Test them!
- Find a construction that seems to maximize what you WANT.
- Find a convincing argument.
- Make sure you check any conjectures before trying to justify them.

Extend

➤ Try maximizing the occurrence of other angles.

➤ What is the maximum number of right angles on the faces of a tetrahedron?

➤ Try polyhedra with as many faces meeting at right angles as possible.

Rolling Coins

Place two coins of the same size flat on a table and roll one around the edge of the other, as if they were gears. When the rolling coin has made one trip around the circumference of the fixed coin, how many times will it have revolved around its own centre?

Entry

➤ Guess first before you try it. Surprised?

➤ Spirograph pieces help accurate specializing.

➤ Be clear about what you WANT.

Attack

➤ Break up the motion in some way.

➤ Could both coins rotate to give the same effect?

➤ Are angles any help?

Extend

➤ What if the rolling coin is one-half, one-third or twice the diameter of the fixed coin?

➤ What if the rolling coin follows more elaborate tracks such as a square, the inside of a large circle, or a figure of eight?

➤ Compare with the motion of the moon around the earth.

Sequence

Write down a sequence of 0s and 1s. Underneath each consecutive pair write a 0 if they are the same and a 1 if not. Repeat this process until you are left with a single digit. Can you predict what the final digit will be?

Entry

➤ Specialize systematically.

➤ Allow your system to alter as you begin to see what is going on.

➤ Try to be systematic about patterns and not about lengths of sequences.

Attack

➤ Try working backwards from the final digit.

➤ Find a convincing argument to support your conjecture.

Extend

➤ Write down a sequence of 0s and 1s in a circle and proceed as before.
➤ Set your result in a more general context by using 0, 1 and 2 with some appropriate rule.

Shadows

I claim that I possess a single loop of wire which, when held up to the sun in three mutually perpendicular positions, casts a square shadow each time. Am I telling the truth?

Entry

➤ Clarify and specialize what you KNOW.
➤ Reformulate more precisely.

Attack

➤ What objects do cast square shadows?

Extend

➤ Can a single loop of wire cast three circular shadows?
➤ What shapes can appear as shadows of a loop of wire? A suitable notation is probably essential here!
➤ What solid objects cast a circular shadow from every direction? Careful! Try two dimensions!
➤ What solid objects cast shadows which always have the same area no matter what the direction? Try it in two dimensions!

Speed Trap

It is rumoured in some countries that the police will not stop you for speeding unless you are going at least 10% over the limit. One such country recently changed from miles to kilometres on all road signs. What is the new rule of thumb?

Entry

➤ Specialize.

Attack

➤ Make a conjecture!
➤ But why? Will it always be so?

Extend

➤ Look for other instances of the role of percentages.
➤ What happens in a country that converts volume measure from gallons to litres, if price rises are usually one or two pence per unit of volume?

Square Bashing

Take any numbers satisfying a pattern of the form

$$4^2 + 5^2 + 6^2 = 2^2 + 3^2 + 8^2$$

Pair up the left and right numbers in any way at all, for example 42, 53, 68. Notice then that

$$42^2 + 53^2 + 68^2 = 24^2 + 35^2 + 86^2$$

Why?

Entry

➤ Specialize.
➤ What do you know about the numbers on the left and on the right sides of the last equation?

Attack

➤ How is what you KNOW related to what you WANT?
➤ Have you tried place value?
➤ Express what you WANT in symbolic fashion.

Extend

➤ Have you tried other pairings?
➤ Will it work for any of these?

$$1^2 + 4^2 + 6^2 + 7^2 = 2^2 + 3^2 + 5^2 + 8^2$$

$$1 + 4 + 6 + 7 = 2 + 3 + 5 + 8$$

$$3^3 + 4^3 + 5^3 = 0^3 + 0^3 + 6^3$$

Square Take-away

Take a rectangular piece of paper and remove from it the largest possible square. Repeat the process with the left-over rectangle. What different things can happen. Can you predict when they will happen?

Entry

➤ Try various proportions. Squared paper might help.
➤ Try using only pencil and squared paper.
➤ Seek a helpful notation.

Attack

➤ Specialize the question. It is too general to start with.
➤ Note how many times the same sized square is removed.
➤ Look for a pattern.
➤ Do you recall finding greatest common divisors?

Extend

➤ Cubes removed from cuboids?

> ## *Sticky Angles*
> Given a supply of sticks, all the same length, and a supply of angles all the same size, can you join the sticks together end to end at the given angle to make a closed ring?

Entry

➤ Have you tried it physically?

➤ Introduce a way of making a supply of angles!

Attack

➤ Have you stayed in the plane?

➤ Have you tried folding a strip of paper appropriately to mimic many sticks joined at the correct angle?

➤ Will your method always work, or is your angle special in some way?

Extend

➤ What is the shortest such sequence when it is possible?

➤ What length sequences are possible?

➤ Does it help to have more than one angle available, particularly when you are confined to the plane?

> ## *Sums of Squares*
> Notice that
> $$2^2 + 3^2 + 6^2 = 7^2$$
> $$3^2 + 4^2 + 12^2 = 13^2$$
> $$4^2 + 5^2 + 20^2 = 21^2$$
> Is this part of a general pattern? Notice also that
> $$3^2 + 4^2 = 5^2$$
> $$10^2 + 11^2 + 12^2 = 13^2 + 14^2$$
> $$21^2 + 22^2 + 23^2 + 24^2 = 25^2 + 26^2 + 27^2$$
> Is this part of a general pattern?

Entry

➤ Can you construct 'the' next example?

Attack

➤ What is the general form of the examples?

➤ Try to describe them without reference to particular numbers.

Extend

➤ What if the first pattern is altered to have three consecutive squares together with one other square?

Taking Matches

Two piles of matches are on a table. A player can remove a match from either pile or a match from both piles. The player who takes the last match loses. If there are two players, how should you play?

Entry

➤ Play the game with someone!
➤ What would be a useful notation?

Attack

➤ Which positions guarantee a win?
➤ How can you arrange to stay in winning positions?

Extend

➤ What happens if there are more than two piles of matches?
➤ What if there are more than two players?
➤ What if the number you can remove is altered?
➤ What if the player who takes the last match wins?

Review

➤ Compare with *Cartesian Chase* (p. 148).

Tethered Goat (silo version)

A goat is tethered to the edge of a circular silo in a grassy field by a rope which reaches just halfway round the silo. How much grass can the goat reach?

Entry

➤ Draw a diagram.
➤ Draw a more accurate diagram.
➤ Try it with a piece of string.

Attack

➤ Replace the circular silo by a figure that looks more like a rectangle (which you KNOW you can cope with).
➤ Try more general figures that look increasingly like a circle.

Extend

➤ Tethered goat problems are easy to make up but not always so easy to solve.
➤ For example, how long must the rope be if a goat tethered to the edge of a circular field is to be able to graze exactly half of the field? Do not expect a simple answer!

➤ What if the goat is tethered to a ring on a wire stretched between two posts? (Idea from Eva Knoll.)

Thirty-one

Two players alternately name a number from 1, 2, 3, 4 or 5. The first player to bring the combined total of all the numbers announced to 31 wins. What is the best number to announce if you go first?

Entry
➤ Try playing it!
➤ What do you need to record?

Attack
➤ What totals enable you to win in one move? Generalize!
➤ Can you find a strategy for playing that will guarantee you to win?

Extend
➤ What if 31 is changed to some other number?
➤ What if the permitted numbers are 1, 2, 3, 4, 5 and 6?
➤ What if there are three players?
➤ What if the permitted numbers are 1, 3, 5 or 2, 3, 7?

Review
➤ Compare with *Taking Matches* and *Cartesian Chase*.

Triangular Count

How many equilateral triangles are there on an eight-fold triangular grid?

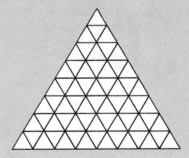

Entry
➤ Specialize? Be systematic!

Attack
➤ Recall a similar question.

Extend
➤ Try larger grids.
➤ Try other grid shapes, like hexagons.

Wool Winding

Wool bought in bulk for a knitting machine comes on cones. The machine uses several cones at one time but, once set up, it is tedious to have to replace an empty cone, and unpleasant to have to rewind wool on to an empty cone. If I have c cones with varying weights of wool on them (all the same colour), and if my pattern calls for k cones at once on the machine, what calculation will predict whether I can use all the wool on the cones without rewinding?

Entry
➤ Specialize. Simplify.
➤ You WANT to calculate with the weights of wool and predict . . .

Attack
➤ Find a condition which must hold.
➤ Find a way of using k cones at a time which uses up all the wool when your condition holds.
➤ Try to reduce what you WANT to a simpler case that you already KNOW.

Extend
➤ Is there a simple strategy to maximize the amount of wool used on c cones in a pattern requiring k cones at a time, without rewinding a cone?

References

Conway, J. and Guy, R. (1996) *The Book of Numbers*. New York: Copernicus, Springer-Verlag.

Fielker, D. (1981) Removing the shackles of Euclid. *Mathematics Teaching* 96, 24–8.

Jaworski, J., Mason, J. and Slomson, A. (1975) *Chez Angelique: The Late Night Problem Book*. Milton Keynes: Chez Angelique Publications.

Moessner, A. (1952) Ein Bemerkung über die Potenzen der natürlichen Zahlen. S.–B. Math.-Nat. Kl. Bayer. Akad. Wiss., **29**(14), 353b.

Noelting, G. (1980) The development of proportional reasoning and the ratio concept part I: differentiation of stages. *Educational Studies in Mathematics*, **11**(2), 217–53.

Streefland, L. (1991) *Fractions in Realistic Mathematics Education: A Paradigm of Developmental Research*. Dordrecht: Kluwer.

11

Thinking mathematically in curriculum topics

This chapter provides a bank of questions that can be used to build a bridge between the highly accessible questions posed and analysed in the main text and topic-based progress in formal mathematics. My aim is to illustrate how the spirit of thinking mathematically can imbue standard topics. Most of the questions here have emerged from contemplating curriculum topics and how learners might encounter them most effectively.

Because the first edition of this book was used in a variety of contexts and educational sectors

- to extend and challenge senior high school students;
- to challenge pre-service primary teachers;
- to challenge pre-service secondary students;
- to challenge undergraduates

this chapter has been arranged by content topics. Some of these are strongly represented in the questions in the earlier chapters. We have also added some topics that are central to the study of upper secondary and undergraduate mathematics, but which were not accessible to a sufficiently wide audience to be included in the main text. However, the division into topic areas is somewhat arbitrary. Many questions use mathematics drawn from different branches of mathematics or that can be approached in different ways, so questions classified under one topic may also fit elsewhere. Most of these questions can be extended or varied so as to challenge people at any level or with any degree of mathematical maturity.

It seems reasonable to assume that anyone dipping into this or the previous chapter is already imbued with advice about specializing in order to regeneralize for oneself, and conjecturing, justifying and reflecting on the actions taken and on their effects. As a result, this standard advice is omitted. There are specific suggestions where these may be needed, which draw attention to curriculum topics and mathematical themes likely to be encountered.

The questions in this chapter are allocated to the following headings. As the depth of investigation of most questions can vary, the education levels are also only indicative:

Place value and arithmetic algorithms
Factors and primes
Fractions
Ratio and proportion
Percentages and rates
Perimeter, area and volume
Number patterns
Equation solving
Geometrical reasoning
Graphs
Functions
Calculus
Sequences and iterations
Mathematical induction
Abstract algebra
Reasoning

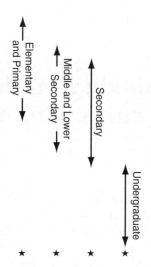

The new questions in the Reasoning section span a broad range of mathematical sophistication.

Place value and arithmetic algorithms

Questions involving digits provide a useful context for developing appreciation of the importance of base-10 arithmetic and algorithms, and a domain for learning to express generality. Thinking of a multi-digit number in terms of its place value, so $234 = 2 \times 100 + 3 \times 10 + 4$, is essential when working on questions concerning digits in base-10 numerals.

Questions from earlier chapters

Palindromes	Chapters 1, 5	Place value and base-10.
Divisibility	Chapter 10	Role of place value and base-10 for various rules of thumb.
Finger Multiplication	Chapter 10	Base-10 properties; algebra explanation uses interesting but elementary expansion.
Cycling Digits	Chapter 10	Place value and base-10; powers of 10.
Reversals	Chapter 10	Place value and base-10; using algebra to express generality.

Additional questions

Copper Plate Multiplication

What is going on here? Describe the 'method' to someone else in writing, so that they can use it on their own multiplications of different numbers of digits.

```
                7   9   6   4   5
                6   4   7   8   9
        ────────────────────────────
                        3   0
                    2   4   2   0
                3   6   1   6   3   5
            5   4   2   4   2   8   4   0
        4   2   3   6   4   2   3   2   4   5
        2   8   6   3   4   8   3   6
            4   9   7   2   5   4
                5   6   8   1
                    6   3
        ────────────────────────────
        5   1   6   0   1   1   9   9   0   5
```

Suggestions

➤ When you have worked it out, make up your own example which shows what to do in every possible situation.

➤ If you can catch the movement of your attention, then you may notice yourself sometimes gazing at some feature (perhaps the whole, perhaps a part); sometimes discerning details not previously noticed; sometimes recognizing relationships between discerned details; sometimes perceiving and treating these relationships as properties that might hold more generally; sometimes reasoning on the basis of those properties to justify your conjectures as to what is going on.

Shifting from 'explaining what is going on' to articulating 'how to do any similar task' is a significant step in understanding and appreciating the power of a technique. A rule becomes a tool when you understand how and why it works.

Grid Locked

What is going on in these ancient calculations?

The first is from an Arabic manuscript *Hindu Reckoning* written by Kushyar ibn-Lebban about 1000 C.E.; the second from the Treviso Arithmetic (1478) and the third from Luca Pacioli (around 1497).

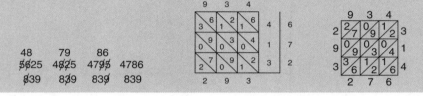

48	79	86	
5625	4825	4795	4786
839	839	839	839

Working out what someone else has done can be an effective way to learn a technique; it certainly makes you think more deeply about arithmetic.

Productive Exchange

$$27 \times 18 - 28 \times 17 = 10$$
$$37 \times 18 - 38 \times 17 = 20$$

Generalize!

Suggestions

➤ 'Say What You See' to yourself; look for variation and invariance. Then try changing one feature and see what happens.

➤ Learning to look for structural relationships rather than simply doing calculations and getting answers contributes to appreciation of the beauties of arithmetic.

Factors and primes

It is an important realization that a number can be named or presented by its digits in base-10 numerals, or by its prime power factors, or indeed in other ways. When looking for patterns, students often look just for additive relationships, but multiplicative relationships often hold the secret. On the face of it, finding the greatest common divisor (highest common factor) of two numbers would appear to require being able to factor the numbers, but

Euclid developed a method which avoids this. Thus gcd (hcf) and the companion lowest common multiple (lcm) mark a significant development in arithmetic and are well worth encountering, even where they are not officially part of the curriculum.

Questions from earlier chapters

Threaded Pins	Chapter 3	Factors; gcd arises because new threads occur when multiples of gap size and number of pins coincide. Compare with *Diagonals of a Rectangle*.
Consecutive Sums	Chapter 4	Using the average of the consecutive numbers helpfully transforms the sum of consecutive numbers to a product.
Square Differences	Chapter 4	Requires turning a difference into a product using a common algebraic identity.
Multi-facets	Chapter 6	Factors; related to *Threaded Pins*.
Diagonals of a Rectangle	Chapter 10	Perhaps unexpected appearance of gcd.
Liouville	Chapter 10	A curious extension of a property of sums of cubes.
Medieval Eggs	Chapter 10	Special case of the Chinese remainder theorem, involving remainders according to different divisors.
More Consecutive Sums	Chapter 10	Depends on how many factors of a particular type.
Odd Divisors	Chapter 10	How factors fit together; closely related to *Porters* (see below).
Square Take-away	Chapter 10	Closely related to Euclid's algorithm for finding gcd (hcf).

Additional questions

Porters

In a certain building there is a long corridor with a very large number of rooms, numbered 1, 2, . . . , and a similarly large number of porters. Each porter k has a key which locks or unlocks each and every door numbered with a multiple of k. The doors all begin locked.

If the porters in turn lock or unlock each of the doors for which their keys work, which doors will be open and which closed? Does it matter in which order they perform their duties?

Given a finite list of open doors, which porters must have been? What properties must such a list have in order to be possible?

Suppose the porters start their ritual with some doors open and some closed. At the end of their ritual, can you, from the state of the doors, work out which doors were open originally?

Suggestions
➤ Be careful not to jump to conclusions; justify your conjectures convincingly!
➤ Closely related to Odd Divisors, but the extensions provide an opportunity to turn the porter's ritual into a mathematical action, and then to study the effects of that action as an object in its own right.

Sieve of Eratosthenes

Write down (use a spreadsheet to print out) the first, say 200 integers in 10 columns. Draw a square around 1. Draw a circle around 2 (the smallest number not yet marked), then cross out every second number starting at 2 (i.e. 4, 6, 8, . . .). Repeat, putting a circle around the smallest number not yet marked (call it m), and crossing out every mth number following it. Continue in this way until all numbers have been marked. What numbers are circled? Why? How many times will a number be crossed out (be careful)? Why? For some numbers, there is no new crossing out to do. Which numbers are these? Why?

Suggestions
➤ When carrying out the process, notice how attention is first only on the process, but then switches to looking at the regularities and seeking reasons.
➤ Observe how the process changes after passing the square root of the end number. Eratosthenes came up with this in the fifth century BCE, and it is

surprisingly efficient. Choosing different numbers of columns emphasizes different patterns. For example, try putting the numbers in just six columns, or 18. Explain why what you see happening must continue.

Remainders of the Day

Find:
- a number that leaves a remainder of 1 on dividing by 2;
- and for which that quotient (the whole number result of dividing by 2) leaves a remainder of 1 on dividing by 3;
- and for which that quotient leaves a remainder of 1 on dividing by 4.

Why must such a number be divisible by 3?

Suggestions
➤ Starting from the beginning and building up not just particular numbers but expressions for all such possible numbers is one approach; starting from the end and building backwards is another approach.
➤ Try varying the remainders involved, looking for other divisibility conclusions. What about longer chains?

Arithmetic is often characterized as working forwards from known to unknown, whereas algebra involves working from the unknown (denoted by a letter or other symbol) towards the known by expressing relationships as equations and solving them. Here there is an opportunity to discover that working backwards can sometimes be more productive than working forwards.

Rational Divisors

Given that 14/15 divides into 28/3 a whole number of times (10 times), we might say that 14/15 is a rational divisor of 28/3.

- Find all the rational divisors of 28/3.
- Find all the rational divisors of 1/2, and then find all the numbers that are rational divisors of both 28/3 and 1/2.

Does it make sense to talk about the greatest common rational divisor of two fractions, and the lowest common rational multiple of two fractions?

Suggestions
➤ Try some simpler cases first, perhaps. The trouble with really simple cases is that they may obscure what happens more generally. Is there a problem arising from the fact that 2/3 and 4/6 are different names for the same rational number?

➤ Making some aspect strange can be effective in bringing to the surface the details of familiar procedures, giving insight into a taste of what it might be like for learners encountering those procedures for the first time.

Chinese Remainders

The set of numbers $\{3 \times 5 \times 3 \times 2 + 3 \times 11 \times 2 \times 2 + 5 \times 11 \times 1 \times 2 + 3 \times 5 \times 11n: n$ an integer$\}$ are exactly the integers that leave a remainder of 2 on dividing by 3, 5 and 11. Why? Generalize.

Suggestions

➤ Try using only two numbers rather than three for finding the remainders. Try changing the 2s to see if you can change the remainders. What then makes it all work?

➤ Remainders work with negative numbers as well as with positives.

Fractions and percentages

Fractions are an obstacle for many learners. The questions proposed here are intended to provide opportunities for exploring with fractions, rather than being an introduction to them. Fraction arithmetic also arises in many questions concerning ratios and rates (see next section).

Questions from earlier chapters

Warehouse	Chapter 1	Expressing percentage changes in terms of multiplication facilitates working with successive changes.
Fractious	Chapter 2	Being clear as to what the 'whole' is, is essential when working with fractions, even when carrying out standard multiplication and division.
Speed Trap	Chapter 10	Expressing a percentage change as a multiplication is useful, as with *Warehouse*. Usefully develops intuition about what changes when measurement units are altered.

Additional questions

Hamburgers

A hamburger is made of bread, tomato, lettuce and meat. If each of the four ingredients increases in price by 5%, by how much does the total cost of the ingredients rise?

Suggestions
➤ Be careful! Explaining the general reason for the right answer may involve more mathematics than expected (including the distributive law). Watch out for people making this slip in the marketplace: it has happened!
➤ This is a very easy problem, but some people get it wrong because their intuition is built on additive structures, not multiplicative ones. Contrast with the additive situation where each of the ingredients increases in price by £2.

Unit Fractions

In how many different ways can $1/n$ be written as the difference of two unit fractions?

Suggestions
➤ There may be more than you expect for some unit fractions. Finding a pattern which generalizes is a good first step, but you need to be sure you have found all possible ways!
➤ Learners are likely to find themselves practising arithmetic with fractions while searching around for all possible ways. A more structural approach is to use algebra to express possibilities and to end up looking at factors of numbers.

Farey Fractions

Take any two fractions. Add the numerators and the denominators to make a new fraction. How does the size of the new fraction compare with the size of the initial fractions? Generalize and explain your result. Find a diagram to display why your result always works (under certain constraints).

Next, make a sequence of fractions by using fraction 1 and 2 to make fraction 3, then using fractions 2 and 3 to make fraction 4, etc. For example, starting with 1/3 and 2/7 we get the sequence 1/3, 2/7, 3/10, 5/17, 8/27, . . .

What happens to this sequence? Does this happen with all starting fractions? Does the result remain the same if fractions that occur in the sequence are replaced by their simplest form (e.g. 5/3, 7/3, 12/6 = 2/1, 9/4, . . .)?

Suggestions

➤ Converting fractions to decimals helps in exploration: decimals are important because they are easier to compare, but working with fractions alone can assist in developing algebraic fluency. Plotting the fractions on a number line assists in understanding what is happening and why.

➤ Adding numerators and denominators is a common error when learners are supposed to be adding fractions. It is however the appropriate operation when adding marks achieved on different questions, for example, but this is not 'fraction addition'. It is appropriate when approximating roots of equations and when combining samples in probability. This question is intended to increase students' awareness of this error by studying the effect of this usually inappropriate operation. Under what circumstances is it appropriate and inappropriate?

John Farey mentioned these fractions in 1816, as did C. Haros in 1802. August Cauchy noticed the mention and proved various things about them. They are closely linked to close-packing of circles all tangential to a straight line!

Uphill

Do there exist four fractions $\dfrac{a_1}{b_1}, \dfrac{c_1}{d_1}, \dfrac{a_2}{b_2}, \dfrac{c_2}{d_2}$ for which $\dfrac{a_1}{b_1} > \dfrac{c_1}{d_1}$ and $\dfrac{a_2}{b_2} > \dfrac{c_2}{d_2}$ but $\dfrac{a_1 + a_2}{b_1 + b_2} < \dfrac{c_1 + c_2}{d_1 + d_2}$?

Suggestions

➤ Trying random examples might not be as helpful as interpreting the fractions as slopes of line segments.

➤ Take any four fractions with different values and order them. Now take the first pair and the last pair.

➤ Consider fractions with the same value as your four but with scaled up numerators and denominators, and try adjusting these. Is there a relation between the differences or the ratios of the combined fraction in terms of the original four?

➤ Consider two different studies of the same phenomenon, leading to your first two inequalities as measures of occurrences of some attribute. Combining the data gives a reversed effect. This is known to statisticians as the Yule–Simpson effect or Simpson's paradox.

Ratios and rates

Questions involving rates, proportions and ratios all belong to the broad class of questions involving multiplicative structures. These questions are studied across nearly all the years of compulsory schooling. Multiplicative structures often behave in counter-intuitive ways, as is highlighted in several of the questions below.

Questions from earlier chapters

Folding Polygons	Chapter 10	Similarity is at the core; ratios, angles and even Pythagoras's theorem and some algebra may come into play.
Fred and Frank	Chapter 10	Distance-time graphs are useful when working with speeds.
Recipes	Chapter 10	Sometimes ratios can be compared using ordinary subtraction, similar to Euclid's algorithm.

Additional questions

Exponential Percentages

- If a population grows by 10% of its current size each month, how long will it take to double its size?
- If a population shrinks by 10% of its current size each month, how long will it take to halve its size?
- If a population alternately grows and shrinks by 10% each month, what happens in the long run?

Suggestions
➤ Does it matter how large the population is to start with? Work out a method for resolving a general class of questions of which this is a particular case. These are examples of exponential growth and decay. What other circumstances have similar growth/decay patterns?
➤ This uses the same observation about a percentage change on a percentage change as is used to resolve *Warehouse*.

Rates

What is the same and what is different about these questions, all of which have caught people out at some time:

1 An old, broken-down car has to travel a two-mile route, up and down a hill. Because it is so old, the car travels the first mile (the ascent) at an average of 15 mph. How fast must it go so as to achieve 30 mph for the whole journey?

2 Two cyclists, initially 30 miles apart, travel towards each other. Cyclist A goes at 14 mph, and cyclist B at 16 mph. A fly flies back and forth between their noses at 30 mph. How far does the fly fly?

3 A trail winds up a mountain and a hiker starting at 6 am at the bottom makes it to the top at 6 pm. On the following day he sets off some time after 6 am and reaches the bottom some time before 6 pm. Is there necessarily somewhere on the path that he reaches at exactly the same time of day both days?

4 A plane cruises at 100 mph in still air. A pilot takes off from A and flies directly to B, 100 miles away, into a 50 mph headwind. Her speed over the ground is therefore 50 mph. She then returns, aided by the tailwind, giving a ground speed of 150 mph. What is her average speed for the whole trip?

5 A man arranges to be picked up at a station at 3 pm but gets an earlier train and arrives at 2 pm. He starts walking along the route, is met and gets home 20 minutes earlier than anticipated. For how long did he walk?

6 Winnie-the-Pooh and Piglet went to visit each other. They started at the same time and walked along the same path. However, Pooh was absorbed in a new 'hum' and Piglet was counting birds overhead so they walked right past each other without noticing. One minute after they met, Pooh was at Piglet's and three minutes after that Piglet reached Pooh's. How long had they each walked?

Suggestions

➤ Don't jump to conclusions! Draw a distance–time graph or otherwise present the relationships to yourself.

➤ Rates, which are one form of named-ratios (others include density, consumption and pressure) cause great difficulty for many people. Attention to the multiplicative relationship and use of mental imagery to 'enter' the situation can be of assistance.

Average Speed

Driving through roadworks on a motorway, the sign said 'speed limit 50; average speed calculated'. I noticed that for a certain number of minutes I was going at 60 mph. For how long do I have to go at 30 to be legal? At 35? Generalize.

I noticed that for a certain distance I was going at 60 mph. How far do I have to go at 30? At 35? Generalize.

Suggestions

➤ Graphs can be a great help, as can entering the situation in imagination and expressing relationships between the entities involved.

➤ Working with rates such as speed is an excellent context in which to develop a deep appreciation of multiplicative relationships.

Cistern Filling

Three spouts direct water into a cistern. They can each fill the cistern in three, four and five days respectively. How long will it take with all three working together?

Suggestions

➤ Try just two spouts first perhaps. Find at least two different ways of thinking about it. Perhaps you can find a way of thinking that leads immediately to expressing the solution, in general. What if you were told the rates of filling when pairs of spouts were working together?

➤ Cistern-filling questions were very popular in late medieval arithmetic books. They provide yet another context in which to work on multiplicative relationships. The reciprocal of the sum of reciprocals is sometimes called *the parallel sum* or *the harmonic sum*.

What is the same and what different about this question and *Arithmagons* or *Weighing Fish*?

Work Force

A man can complete a certain task in three hours, a woman in four and a child in five. How long will it take them working together?

Suggestions

➤ Vary the task to include multiple men, women and children. What if you are told the rates of working in pairs rather than singly?

➤ What is the same and what different about cistern filling and people working?

➤ What other variants can you come up with?

Questions like this with different rates of work were common in Victorian arithmetic books. Their use provides us now with informative cultural commentary; then, the questions provided yet more experience of multiplicative relationships and harmonic sums.

Outing

People leave two towns at the same time, some going in each direction, and they all meet at noon where they have a meal together for an hour. One group continues and reaches the other town at 7:15 pm, while the other group gets to the first town at 5 pm. When did they start?

Originated by Arnold and transmitted to us by a chain of people ending with Peter Liljedahl.

Suggestions
➤ Draw a diagram or graph to display distance against time.

Ride and Tie

In the nineteenth century before the advent of motor vehicles, people would sometimes share a horse when making a long journey. One person would ride while the other walked; the rider would tie the horse at a convenient point and walk on, while the other would pick up the horse and ride on. The horse could rest whilst waiting for the second rider. This action might be repeated several times. How must the ride and tie options be adjusted so that the two travellers arrive at their destination at the same time?

Suggestions
➤ A graph is certainly helpful. It is conventional to put the time on the horizontal axis and the distance from the start on the vertical axis. Deciding what information you need and assigning letters to it also frees up thinking. Using dynamic geometry is even more helpful, since you can then adjust the places where the horse is tied up waiting for the other traveller. You might then notice an invariant, and so discover exactly what information you really need in order to arrange to arrive at the same time.

Ride and Tie is described by Henry Fielding (*History of Joseph Andrews and his friend Mr Abraham Adams 1742*) and mentioned by Thomas Paine in his *Rights of Man* (p. 33). In 1798 James Carnahan and Jacob Lindley used this method to cross the Alleghany Mountains to Princeton University to start their undergraduate careers; Carnahan later became president of Princeton, and Lindley president of Ohio University. *Ride and Tie* has become the basis for sporting events. There are requirements that the horse be allowed to rest for a specified time each hour. Of course you could introduce further means of transport and more travellers, using a bicycle, scooter, etc.

The ride and tie problem was posed by Ransom (1952) and a solution printed in Ransom and Braun (1953). Mention is made of several men and one horse, and that several riders and several horses is a 'much more involved problem'.

Newton Grazing

If 12 cows fully graze a pasture of 10/3 acres in four weeks, including all the grass that grows during that time, and 21 cows fully graze 10 acres in nine weeks, how many cows will fully graze 36 acres in 18 weeks?

Suggestions
➤ The role of *fully grazing* is important here. At the end of the time there is no grass left for grazing, but in the meantime, the grass grows while being grazed.
➤ It might help to work out how long it will take one cow to graze one acre, or how many acres one cow can graze in just one day.

Isaac Newton posed this question in his text on algebra, solving it both in particular and in general. You can imagine him gazing out of his window and seeing cows grazing on the nearby meadow. Newton's collection of questions marked the end of his interest in solving 'word problems'. Of greater interest to mathematicians was how to solve the equations that arise from expressing word problem relationships algebraically.

Propensities

In a certain club, 10% of the members were poets. Some members were to be selected to perform at an event, and 40% of these performers were to be poets. Explain why poets are six times more likely to be selected than non-poets.

If the numbers of objects of a given colour in a collection are in the ratio of $c_1 : c_2 : \ldots : c_n$, and if those with a certain characteristic P are in the ratio $p_1 : p_2 : \ldots : p_n$ then what are the ratios of the occurrences of different colours having characteristic P?

Suggestions
➤ Surprised? It pays to be very clear about exactly what you are trying to find out!
➤ Try using a rectangle to specify the membership, and dividing the rectangle by horizontal line and a vertical line to represent the two selection processes (poets and performing).
➤ Moving to more than two 'colours' may require confidence with the calculation in the special case of two 'colours'.
➤ Generalizing can sometimes bring about a change of perspective on the original problem.

Equations

Solving equations by algebra is one of the most successful tools for attacking mathematical questions. Isaac Newton is one of the mathematicians who, seeing how effective it was, turned his attention and that of other mathematicians to the problems of solving the equations that you end up with and away from pedagogical problems of expressing the relationships in algebraic terms. As can be seen in several of the problems below, using algebra often turns a puzzle into a routine exercise. This is the power of algebra. However, solving routinely also divorces the problem solving from the problem context, and then there is a danger that meaning is lost.

Questions from earlier chapters

Creepy Crawlies	Chapter 2	This is an example of a Diophantine equation (one which must be solved with integers). This additional information enables the solution of some sets of equations with seemingly too little information.
Rolling Coins	Chapters 2 and 3	Polar coordinates can assist.
Eureka Sequences	Chapter 5	Conjecturing and using counter-examples.
Arithmagons	Chapter 10	Excellent to compare the solutions by logical arithmetic reasoning and by solving equations.
Half Life	Chapter 10	Expressing what is not known with symbols can be very helpful, but requires care. Consider if the variable stands for the person's age at one instant in time, or at every time. Highlights the importance of being precise about what quantity is represented by the letter.
Leap Birthday	Chapter 10	Setting up and solving equations is simple, but interpreting the answer is difficult.
One Sum	Chapter 10	Algebra encapsulates the properties of number operations, and routine manipulation enables them to be marshalled together to make a routine solution.

Additional questions

Weighing Fish

A fisherman caught three fish. The fish were not weighed separately, but instead all the pairs of two fish were weighed together. The big fish and the middle-sized fish together weighed 16 kg. The big fish and the small fish together weighed 14 kg. The small fish and the middle-sized fish together weighed 12 kg. How much did each fish weigh?

Suggestions

➤ Solve this by setting up three linear equations in three unknowns. Then try to solve it without equations – it is not too hard. Then see how the two solutions correspond. What insights do you get from each method? Compare your methods with *Arithmagons*.

➤ Next, generalize: what if the weights of the fish are different? Relax the condition that the weights of fish have to be positive. How does your equation method and the solution by logical reasoning generalize? Next, generalize to more fish, still weighing all the pairs. There are several different types of solutions, depending on the number of fish. What if you weigh combinations of three or more fish instead?

Like *Arithmagons*, *Weighing Fish* deserves to be generalized and studied carefully. A great deal of linear algebra could be taught through these questions. The theory of linear equations in many variables explains the various behaviours exhibited when the number of fish increases.

Age Problems

The combined ages of *A* and *B* are 48 years, and *A* is twice as old as *B* was when *A* was half as old as *B* will be when *B* is three times as old as *A* was when *A* was three times as old as *B* was then. How old is *B*?

Suggestions

➤ An empty number-line can be used to keep track of the information at different times; algebra is very handy.

➤ It is at least as much fun to make up your own variations, and at the same time this gives insight into the internal structure of such problems.

Moving Mean

In a list of positive integers, the arithmetic mean is 5, and the number 16 is known to appear. If the 16 is removed, the mean drops to 4. What is the largest possible number that could occur in the original list, and how many numbers were in that list?

Suggestions

➤ Clarity about what is known and what is wanted could be really helpful. Of interest is to work out how to produce more examples like this. A good way to appreciate a construct like arithmetic mean is to work with it in a multitude of ways.

Patterns and algebra

In some of these questions, the main challenge is to describe a pattern algebraically, and then prove it by algebraic manipulation. In other questions, the main challenge is to gather numerical evidence, discover a pattern, describe the pattern algebraically, and then explain why the described pattern correctly describes the mathematical structure.

Questions from earlier chapters

Chessboard Squares	Chapter 1	Finding a way to count systematically, and to express a general polynomial relationship.
Paper Strip	Chapter 1	Counting by recognizing recursive relationships; exponential formulas involved.
Chessboard Rectangles	Chapter 2	Extending a technique to a new setting; making use of previous generality to express a greater one. Polynomial relationships.

(continued)

Leapfrogs	Chapter 3	Acting systematically; recognizing relationships; expressing these as generalities. Provides a reason to generalize before encountering algebraic manipulation. Squares and products involved. Compare result with *Perforations*.
Circle and Spots	Chapters 4, 5	Not jumping to conclusions about generalities on the basis of scant evidence, then finding deeper relationships. Binomial coefficients and their sums.
Bee Genealogy	Chapter 5	Importance of relating patterns in numbers arising from a situation, with structural features of that situation. Fibonacci numbers.
Matches 1 and 2	Chapter 5	Linking structural aspects of a physical situation with those of a numerical relationship; elementary algebra.
Number Spirals	Chapter 8	Linking structural aspects of the source and sequences of numbers that arise; conjecturing and also convincing people it will always happen.
Handshakes	Chapter 10	Recognizing and expressing structural relationships, involving sums of consecutive numbers or a product.
Polygonal Numbers	Chapter 10	Linking structural relationships in the geometrical layout with those in the associated number sequence.
Right Angles	Chapter 10	Relating geometric implications of right-angles with counting. Some patterns observed might be false. Generalization requires greatest-integer-less-than function.
Sums of Squares	Chapter 10	Detecting and expressing relationships between numbers; generalizing and convincing; algebraic manipulation.

Additional questions

Paper Strip Expressions

Review how the folding is done in *Paper Strip* (Chapter 1). Let $C(n)$ be the number of creases in the strip after n folds and let $S(n)$ be the number of sections that the strip has been divided into after n folds. Explain why each of the following are true:

$$S(n + 1) = 2S(n) \qquad\qquad C(n) = 1 + 2 + 2^2 + 2^3 + \cdots + 2^{n-1}$$
$$C(n + 1) = C(n) + S(n) \qquad C(n) = 2^n - 1$$
$$C(n) = S(n) - 1$$

What other true statements are there involving $C(n)$ and $S(n)$?

Suggestions

➤ One approach is to make a table of values for n, $S(n)$ and $C(n)$, observe patterns in the table and interpret the statements above in terms of this table. This leaves open justification for the patterns in the table to be valid in terms of the original context of paper folding. Another approach is to justify the statements in terms of the action of folding, and then deducing the tabled values.

➤ The fact that $1 + 2 + 2^2 + 2^3 + \cdots + 2^{n-1} = 2^n - 1$ can be shown using the formula for the sum of a geometric series, but it can also be shown by arguing in terms of the number of sections and creases and using the relationships above. Algebra is an abstract language which does not reference the context directly, but insight into algebra can come by thinking about concrete contexts such as this one.

Perforations

In the days when sheets of stamps were perforated to make it easy to detach one at a time, a sheet of six stamps would look something like the picture. How many perforations would there be for a sheet of r rows and c columns of stamps?

Generalize to w perforations on the width and b perforations on the breadth, and c perforations in the corners.

If someone announced they had a sheet with a specified number of perforations, how could you check whether that was possible (and in how many ways) without actually constructing sheets of perforations?

Suggestions
➤ Some people find it helpful to fix one or more parameters so they can explore effects in changing the others.
➤ Curious things happen when you substitute in 0 for the number of rows and the number of columns of stamps!
➤ To make sense of the 'undoing' extension, find some way to manipulate your expression so that it can be expressed in a more perspicuous fashion, for example as a product. Then apply that product requirement to the correspondingly manipulated specified number of perforations.
➤ In its full generality this can be a challenging problem. It is likely to bring to the fore the systematic control of variation, and paying attention to how you draw examples as a guide to how the perforations can be counted.

Rooted

Generalize the following observations:

$$2\sqrt{\frac{2}{3}} = \sqrt{2\frac{2}{3}} \qquad 5\sqrt{\frac{5}{24}} = \sqrt{5\frac{5}{24}} \qquad 6\sqrt{\frac{6}{35}} = \sqrt{6\frac{6}{35}}$$

What is the range of permissible change of your variable?

Suggestions
➤ What is the same and what is different? What is invariant and what is changing?
➤ Some learners may be prone to making 'mistakes' that look like this. Here attention is drawn to the hidden plus sign in mixed fractions and the hidden multiplication outside the square root sign.

Divisive Subtractions

Generalize the following observations:

$$4 - 2 = \frac{4}{2} = \frac{2}{1} \qquad \frac{16}{3} - 4 = \frac{16}{12} = \frac{4}{3} \qquad \frac{49}{6} - 7 = \frac{49}{42} = \frac{7}{6}$$

What is the permissible range of change?

Suggestions
➤ What is the same and what is different? What is invariant and what is changing?
➤ Use relationships evident in the second and third to inform a rewriting of the first, and express this as a general characterizing or generating property.

Working on the patterned relationships that emerge provides plenty of practice in subtracting fractions as well as expressing generality.

Cube Result

Someone noticed that $2 \times 3 \times 4 + 4 \times 10 = 4^3$ and that $5 \times 6 \times 7 + 7 \times 19 = 8^3$. Are these examples of some general pattern, or are they simply anomalies and not part of a continuing pattern?

Suggestions

➤ What is the same and what is changing, and in what way? What relationships might be present?

➤ Does your generality work for negative numbers? What about for fractions?

➤ Might there be something similar for fourth powers?

➤ Discerning specific elements, recognizing relationships between them and expressing these as properties together constitute a structural approach. Then it may be wise to check on other examples. Generating more examples and then looking for relationships is a more empirical approach.

Graphs and functions

Graphs are often seen as the end point of mathematical work, rather than as a way to present relationships visually, which can then be interpreted or explored further. Being able to view relationships both algebraically and graphically provides great power. Work within the curriculum often focuses exclusively on graphs of functions, but graphs of relations that are not functions are also interesting.

Additional questions

Differing By Two

● Sketch the graphs of two straight lines whose slopes differ by 2. And another pair; and another pair.

● Sketch the graphs of two straight lines whose x-intercepts differ by 2; and another pair; and another pair.

● Sketch the graphs of two straight lines whose y-intercepts differ by 2; and another pair; and another pair.

● Now sketch the graphs of two straight lines whose slopes, x-intercepts and y-intercepts all differ by 2.

Write down expressions for all such pairs. What is special about the 2?

Extend.

Watson and Mason (2006)

Suggestions

➤ Need the 2s be the same for each aspect of a pair of straight lines? What can you do in three dimensions?

➤ The idea behind 'and another' is that by the third instruction many people are starting to look for more interesting examples, which provides a stepping stone to expressing generality. The questions have been structured so as to build up constraints. By constructing the general class of objects at each stage, each additional constraint can be coped with more easily. When I meet a construction task with several constraints, the option is open to impose them sequentially rather than all at once.

Rotations

Under what conditions can you rotate the graph of a function about the origin, and still have the resulting graph being the graph of a function? If the graph of a function cannot be rotated about the origin without ceasing to be the graph of a function, might there be other points which could act as centre of rotation and preserve the property of being the graph of a function?

Suggestions

➤ Try some familiar functions, in order to see what quality either enables or blocks a rotation. Start with particular angles of rotation, such as 180 degrees and 90 degrees.

➤ Work on this task can contribute to awareness about what happens to functions as x gets very large both positively and negatively. Also available is the difference between a graph as an object and the function of which it is the graph, seen as a set of points. The concept of slope may also be helpful.

Functioning Oddly and Evenly

If a function f is invariant under reflection in the line $y = 0$, then it is referred to as an even function. If it is invariant when reflected twice, once in the line $x = 0$ and once in the line $y = 0$, then it is referred to as an odd function. Which functions can be written as the sum of an even and an odd function?

Suggestions

➤ Express what it means to be even (odd) symbolically as a property of $f(x)$.

➤ Explore the effect of composing a function with an even function or an odd function, and try composing two odd functions.

Reflections

If a function is one-to-one on some domain, then its reflection in the line $y = x$ is also a function (the inverse function) on a corresponding domain. Are there any functions which are symmetric in the line $y = mx$ for some m? For each m, classify the functions for which the reflection in the line $y = mx$ is also a function.

Suggestions
➤ Work out how to reflect any point in the line $y = mx$. Then require that the image of the set of points $[x, f(x)]$ be the set of points on a function. It may help at each stage to parallel the computation with the special case when $m = 1$. Drawing graphs on transparent plastic could make them easier to manipulate.

This question shows the special nature of one-to-one as a property that guarantees the existence of an inverse function by placing it in a more general context.

Properties of Polynomials

For any polynomial and any interval I form the chord AB of that polynomial on the interval I. According to the mean value theorem, there is at least one point on I where the tangent is parallel to the chord. Where would you look for such a point?

In other words, classify the curves for which the point of tangency corresponds to the mid-point of the interval for every interval. Is there a polynomial for which there is a fixed ratio other than 1:1 at which the tangent occurs? What happens as the interval width gets smaller and smaller?

Suggestions
➤ Experiment with familiar functions. Find a way to present functions making use of the derivatives so that you can see what happens as the interval gets smaller.
➤ Finding a relevant property or theorem to use is not always straightforward. Taylor's theorem may be of assistance.

Properties of Cubics

Imagine the graph of a cubic polynomial. Draw a chord between two points on the cubic. Mark the mid-point of the chord and draw the vertical line through the mid-point. Now take any other chord whose mid-point is on the same vertical line. Where do the two chords intersect?

Suggestions
➤ Dynamic geometry software can be of considerable value but the images cannot be believed unless and until the conjectured property is justified convincingly.
➤ If the cubic has three real roots, what happens if you start with the chord through two of the roots? Extend to the case where the first chord is parallel to the x-axis.

Exploring properties of polynomials generates familiarity with them as mathematical objects. This result generalizes in several directions.

Symmetry of Cubics

It is well known that the graphs of all quadratic functions are symmetric. It is not so well known that the graphs of all cubic functions are also symmetric. What sort of symmetry do they all have? Prove your answer.

Suggestions
➤ A good way to start is by specializing, perhaps printing some graphs on transparent plastic and experimenting. Identify geometric/spatial features of the graphs of cubics, and explore those symmetries that would preserve those features. Using a computer algebra system can take the hard work out of the calculations for a proof.

Questions like this give students an opportunity to get to know classes of examples thoroughly (in this case cubics), which makes them more likely to call upon them as examples when they meet other concepts.

Chordal Divisions

● What is the set of points which can be the mid-point of a chord of a given quadratic function?
● What if mid-point is replaced by some other ratio? What if that ratio is greater than 1 or less than 0?

Suggestions
➤ Two strategies are available. They approach the question in opposite ways. Fix one end of a chord and let the other end move along the curve, or fix the x-coordinate of the mid-point, and let both ends of the chord vary.
➤ Extending to cubics is interesting; extending to quartics is surprising but obscure.

Functions and calculus

The concepts and tools of calculus make up a major component of senior secondary and undergraduate mathematics. The following questions focus attention on central concepts in new ways, and apply calculus ideas in unusual geometric contexts.

Additional questions

Tangent Power

Given a smooth function f (say twice differentiable) on R, define the *tangent power* of a point P to be the number of tangents to f through P. Investigate the regions of the plane which have the same tangent power.

Suggestions

➤ You can start with a point and imagine the tangents through it, or you can imagine a tangent moving along the curve and swinging about through the plane.

This task is useful for developing an awareness of the graphs of functions as x gets very large in absolute value. It also provides an introduction to the second derivative.

Slippery Slopes

Draw the graph of a smooth function on R.

● Fix a point P and sketch the graph of the slope of the chords from P to points Q on the function, against the x-coordinates of Q. What happens as Q gets close to P?
● Fix an interval width δ and graph the slope of the chord from (x, f(x)) to (x + δ, f(x)), against x. Repeat for smaller values of δ. What happens to the curve as δ goes to zero?
● Fix a radius r and graph the slope of the chord from (x, f(x)) to (t, f(t)), against x where the distance between these points is r and where t > x (this is difficult to do in general!). What happens to the graph of the slope as r goes to 0?

Suggestions

➤ Dynamic geometry or computer algebra systems are ideal tools for creating the first two constructions as animations in order to get a dynamic sense of different ways of thinking about slopes of tangents and derivatives. The algebraic obstacles of the last one indicate why this does not feature in defining the slope of a smooth curve.

Chordal Properties of Quadratics

Take any two points A and B on a quadratic and draw the chord AB. Let M be the mid-point of AB. Let C be the point on the quadratic vertically in line with M. Draw the chords AC and BC. Now form the mid-points M_A of AC and M_B of BC and the corresponding points D and E vertically in line with them on the quadratic. What can be said about the lengths of the segments $M_A D$ and $M_B E$?

Take any chord AB of a quadratic. Draw the tangent to the quadratic which is parallel to AB. Compare the point of tangency to the mid-point.

Suggestions
➤ Try using the simplest possible quadratic. Can the reasoning then be extended to all quadratics? All parabolae (you might have to reinterpret 'vertical')? Finding a suitable way to present the chords and their mid-points is part of the process.

Tangents to Quadratics

What is the locus of points from which the two tangents to a quadratic are at a specified angle?

Suggestions
➤ Algebra can be used to derive equations by means of which to recognize the locus but is there a geometrical way to see the locus? Now apply *doing and undoing*: given such a locus, what is the collection of quadratics and associated tangent angles that give rise to it? Affords experience with coordinates of points on graphs of functions.

Tangents between the Roots of Cubics

Let a and b be any two distinct roots of the cubic function $f(x)$. The graph of $f(x)$ will cross the x-axis at $(a, 0)$ and $(b, 0)$. Let P be the point on the graph of $y = f(x)$ midway between these roots (i.e. with x-coordinate $(a + b)/2$). Construct the tangent to $f(x)$ through P and find where it intersects $f(x)$. Prove the surprising result that you will notice.

Thinking Mathematically

Suggestions

➤ Illustrating this problem with dynamic geometry brings it to life and enables a conjecture to be made quickly. When proving the result, consider the best way to represent $f(x)$, as a sum of terms or as a product of factors. Selecting an algebraic form to highlight the most significant structure for a problem is part of the art of doing mathematics. Because this problem involves the roots of $f(x)$ expressing it as a product is my choice. Provides practice with using coordinates of points on graphs of functions. What is special about the roots? Might any chord serve as well? What about extending to higher degree polynomials, and/or not just chords but parabolae through three points, and beyond.

Integration by Parts

Construct a function that requires two integrations by parts in order to find its indefinite integral.

Suggestions

➤ Reminding yourself that integration by parts comes from the formula for the derivative of a product may be helpful.

This task only gets interesting when you go beyond two to three, four, and beyond, and when you seek out different ways to force integration by parts. At the same time it provides a good exercise in integration by parts and awareness of when it is an appropriate technique to use.

L'Hôpital

Construct a function that requires two uses of L'Hôpital's method (actually due to Johann Bernoulli) for finding limits.

Generalize: change the two; find different ways to force multiple use of the method.

Suggestions

➤ Taking any technique and trying to construct examples which require multiple uses of the technique generates a richer insight into what the technique accomplishes and how it works.

Area Cuts

Archimedes discovered that if you take a chord of a parabola, then the area of the triangle with base the chord and vertex on the parabola corresponding to the mid-point of the chord is $\frac{3}{4}$ of the area between the parabola and the chord and also the maximum area for any triangle in the region with the chord as base.

Are there other functions with the same or different constant?

Suggestions

➤ Find a streamlined calculation to check Archimedes' conjecture. Then try mid-points of chords on cubics before extending further.

➤ For a cubic, consider the areas of the regions between the curve and above and below any line though the inflection point. For a quartic, consider the areas between the curve and above and below the line through the two inflection points. Can you extend to quintics in some way?

➤ This is good practice in setting up and calculating integrals for areas, as well as dealing with several unknowns. Parametric presentation of the points on the curve may make the calculations easier than otherwise.

Compounding Functions

Using the function $f(x) = x^2$, how many different functions can you make using the operations of adding, subtracting and composing functions, using two, three, four, . . . occurrences of f? Try to describe in words how you would recognize such a function.

Using the functions $f(x) = x^2$, $g(x) = x - 1$ and $h(x) = 3x$, together with the operations of adding, subtracting, multiplying and composing, how many different functions can you form using all three functions each time? Try to describe in words how to recognize a function made in this way.

Try using other triples of functions.

Suggestions

➤ Constructing your own examples is more engaging than doing someone else's exercises, and at the same time leads to internalizing and automating of routines: here, composition of functions.

➤ Try using other functions in place of x^2, such as x^3, etc., or $x^{1/2}$, etc.

Sequences and iteration

Detecting and expressing the structure of how a sequence is generated, or iterating a sequence according to a specific rule can lead to excellent opportunities for expressing generality and for making and justifying conjectures. This provides a rich background for later work on limits.

Questions from earlier chapters

Paper Strip	Chapter 1	Relating input sequence to a sequence of effects.
Iterates	Chapter 5	Famously unsolved problem; notion of an iterated action.
Ins and Outs	Chapter 10	Relating output sequence to a sequence of actions. Complex pattern of iteration in an accessible setting.

Additional questions

Cyclic Iterations (A)

The iteration $u_{n+2} = u_{n+1} - u_n$ repeats itself after six iterations, no matter what the two starting numbers are. So does $u_{n+2} = u_{n+1}/u_n$. Experiment with other iterations such as $u_{n+2} = (1 + u_{n+1})/u_n$ to get other cycle lengths.

Suggestions

➤ Try changing the additive constant 1; try changing the sign.

➤ Try imposing on the parameter t conditions to force the iteration $u_{n+2} = t u_{n+1} - u_n$ to have specified cycle length.

The shift from finding cycle length to looking for iterations with different cycle lengths, to imposing conditions on parameters is a common mathematical approach to exploration.

Cyclic Iterations (B)

Pick any non-zero number p. Now select two starting numbers, a and b, and then iteratively apply the action which replaces $[a, b]$ with $\left[b, \dfrac{p(b + p)}{a} \right]$.

How does this relate to *Cyclic Iterations (A)*? Why does it do what it does? Find other iterations with different cycle lengths.

Suggestions
➤ Pick simple but not too simple numbers for p, a and b to start with or, if you are strong algebraically, work with the letters. Conjectures can be checked quickly using a spreadsheet.

Limitations

Can it be true that $3 = \sqrt{1 + 2\sqrt{1 + 3\sqrt{1 + 4\sqrt{1 + \ldots}}}}$?

Jackson and Ramsay (1993); Rabinowitz (1992)

Suggestions
➤ Finding a recursion that is always valid, a method of 'unfolding' to make longer and longer sequences, is certainly supportive, but not fully convincing, mathematically.
➤ Validating the limit if the limit exists is relatively straightforward; showing that the limit exists is another matter!
➤ Make up your own similar sequences by using a similar unfolding and substitution technique.

Unfolding 9s

Let $x = 0.99999\ldots$

Then $x = 0.9 + x/10 = 0.99 + x/100 = 0.999 + x/1000\ldots$ Each of these equations can be used to deduce a value of x. Call this the *method of unfolding* for finding the value of infinite processes if they converge. What is the value of x?

Try applying the unfolding process to $1 - 1 + 1 - 1 + 1\ldots$

Suggestions
➤ Euler decided that a suitable value for this infinite process of adding and subtracting 1 was $\frac{1}{2}$. Why might he have concluded this? Cauchy would not accept this value, because the sequence does not actually converge according to Cauchy's criteria. Instead, the partial sums alternate between 0 and 1.

Unfolding Fibonacci Relationships

Let x be the value of the following infinite process (if it converges):

$$1 + \cfrac{1}{1 + \cfrac{1}{1 + \cfrac{1}{1 + \cfrac{1}{\cdots}}}}$$

Now use the idea of unfolding to replace part of the calculation by itself. For example,

$$x = 1 + \cfrac{1}{x} = 1 + \cfrac{1}{1 + \cfrac{1}{x}} = 1 + \cfrac{1}{1 + \cfrac{1}{1 + \cfrac{1}{x}}}$$

Let y be the value of $\sqrt{1 + \sqrt{1 + \sqrt{1 + \ldots}}}$ (if it converges). What is the value of y?

Let z be the value of $-1 + \left(-1 + \left(-1 + (\ldots)^2\right)^2\right)^2$ (if it converges). What is the value of z?

Let w be the value of
$$1 + \cfrac{1}{1 + \cfrac{1 + \cfrac{1 + \ldots}{(\cdots)}}{1 + \cfrac{1 + \cfrac{1 + \ldots}{(\cdots)}}{1 + \ldots}}}$$
(if it converges).

What is the value of w?

Suggestions

➤ One approach might be to approximate these expressions by cutting them off and evaluating them, or finding what happens as you take more and more terms.

➤ Another approach might be to assume they have meaningful values and then make use of the infinity in their description to formulate a relationship between two expressions using that value.

➤ Explore other sequences like these. Which ones converge?

Provides an opportunity to work on seeing relationships between parts and wholes, and then justifying the conjecture that it does actually converge.

Accumulating Limits

What characterizes those sets S on the real line which can be the set of accumulation points (limits of subsequences) of some subset of the reals?

Suggestions

➤ Start building up examples, and seeing their sets of accumulation points as representing types of subsets of the reals.

See Thomas Sibley (2008) for one approach. By exploring questions about what sorts of objects are possible as the results of a specified action on other objects, you get a much deeper sense of what that action entails.

Mathematical induction

Many of the general formulae obtained in the questions above and elsewhere in this book can also be proved by mathematical induction. Proof by mathematical induction is widely applicable and is often the easiest approach once you have a formula. However, it does not always provide as much insight as a proof that uses the structure within the problem directly. The initial chapters of this book have emphasized structure rather than proof by mathematical induction.

Questions from earlier chapters

Patchwork	Chapter 1	Justifying a general formula formulated through specializing; requires insight into how additional regions are created.
Chessboard Squares	Chapter 1	Justifying a general expression with one unknown.
Chessboard Rectangles	Chapter 2	Justifying a general formula with two parameters (length and width).
Leapfrogs	Chapter 3	Justifying a general formula, with two parameters if different numbers of pegs are allowed on right and left sides.
Circle and Spots	Chapters 4 and 5	Justifying a general formula involving binomial coefficients or, more empirically, a general formula that is obtained by fitting a polynomial to data points.
Polygonal Numbers	Chapter 10	Plenty of opportunity to justify formulae.

Additional questions

Sorting Stacks of Numbers

You come across some stacks of carpet tiles, each of which has a unique number on it. The tiles are quite large and heavy so you can only move one at a time from one stack to another. You pick a positive number d, and you resolve to place a tile on top of another *only if* the tile you are moving is at least d smaller than the number on the tile currently on the top of the stack. If you can end up sorting the tiles into stacks of decreasing numbers bottom to top, you will clearly need at least d stacks. What is the minimum number of stacks needed to guarantee you can always sort the tiles for a given d?

Suggestions

➤ It might be worthwhile specializing to $d = 1$, the 'Tower of Hanoi' version, but here starting with the tiles in some random order on each stack. In order to be certain that you can always move at least one tile, you need $d + 1$ stacks. Is this sufficient?

➤ Induction on the number of tiles might be possible. Think in terms of an action and what it preserves.

➤ Choosing what to induct on is not always as straightforward as it seems.

Chessboard Squares – Mathematical Induction

Prove that the number of squares on a chessboard is equal to $n(n + 1)(2n + 1)/6$. This is a different form of the solution to the one found in Chapter 1.

Is it possible to explain this result in terms of the chessboard, or can it only be shown using algebra to modify the expression derived in Chapter 1?

Suggestions

➤ Might there be a way to arrange physical squares with unit thickness in three dimensions? Perhaps the six can be accounted for by arranging six copies of all the squares. Finding ways to arrange objects so that they can be conveniently counted is part of the art of combinatorial thinking.

Patchwork Completed

Prove the result for *Patchwork* by mathematical induction.

Find out about the history of the Four-colour Theorem. What is the difference in conditions between *Patchwork* and the Four-colour Theorem?

Suggestions

➤ Be sure to cover all cases in the inductive step. Take care to select the variable to induct upon – it has to be a natural number and cannot be a geometric arrangement or other mathematical object. There are many patchwork arrangements with a given number of lines or regions. How will you be sure to have covered all cases?

Abstract algebra

These questions can be used as access to group theory and related topics when extended beyond the specifics, yet the first one is accessible to young children if posed appropriately and can be varied to give multiple exposure to the same basic idea. Together, the problems use, or can develop, ideas that are central to abstract algebra such as closure, binary operation and properties of operations such as associativity.

Questions from earlier chapters

Furniture	Chapter 5	When generalized, involves idea of a group.

Additional questions

Remainder Primes

This problem considers *only* the numbers which leave a remainder of 1 on dividing by 3. Write down 10 of these numbers and check that if you multiply any two of these together you get another number in the set. Why is this?

Write down the first 10 numbers that are 'prime' in this set. These are numbers in the set that cannot be written as a non-trivial product of numbers of that form. What is the same and what is different about factoring 100 into primes in the usual system, and in this system?

What is special about the 1? About the 3? What if they change?

What about considering *only* the numbers that leave a remainder of 1 or of 4 when divided by 5. Do they preserve this property when two of them are multiplied together? Are the primes the same when you consider *only* numbers leaving a remainder of 1 and when you also permit numbers leaving a remainder of 4?

Suggestions

➤ Suitable test-bed examples may not be straightforward to construct since the examples need to be neither too simple nor involve too much arithmetic.

➤ You only really understand a concept (here, *primes*) when you extend it and experience similarities and differences. The *fundamental theorem of arithmetic* is that factoring into primes gives a unique decomposition of the number. But unique factorization is not universal!

This task is also a pathway into the mathematical area known as *group theory*.

A fruitful extension is to consider *Gaussian primes*: complex numbers of the form $a + ib$ where a and b are integers. You can also change i to $\sqrt{5}$, or even $\sqrt{-5}$, or some other surd, and ask about unique factorization. First check that if you multiply two together you remain within the system selected.

König's Theorem

In a rectangular grid, some cells have a counter while others do not. A *cover* is a set of rows and columns which include or cover all of the counters present. An *independent set* of counters is a subset of the counters with no two in the same row or column.

Claim: the smallest number of rows and columns that cover all the counters is equal to the size of the largest possible independent set of those counters. In other words, the size of a minimum cover is the size of a maximal independent set of counters.

A bipartite graph has two distinct sets of vertices, and all edges join a vertex in one set to a vertex in the other.

Claim: the minimum number of vertices meeting all the edges is the maximum number of edges with no vertices in common.

Try proving that each of the two claims can be proved from the other. But are the claims true?

Suggestions

➤ What is needed is a train of reasoning. One may be easier to think about than the other, so proving they are equivalent may be helpful. A few examples may provide access to underlying structure. Using an example to try to relate the two claims is an exercise in interpretation or modelling.

➤ While substantial reasoning is involved, there are no other complicated mathematical ideas required. A good strategy is to find a way to make an independent set of counters larger if all covers are larger (or alternatively, to make a cover smaller if all independent sets are smaller). This demonstrates a fundamental idea of showing that any potential candidate (in this case for the largest set) can be modified given the other information (in this case, that all covers are larger.) What could happen in three dimensions?

Find the Identity!

Take the set of numbers {1, 2, 3, 4} under multiplication modulo 5 (take the remainder on dividing by 5). The product of any two of them is again one of them. Now multiply each number by 6 and this time use multiplication modulo 15 (take the remainder on dividing by 15). Which element is the identity under multiplication modulo 15? Similarly for multiplying each number by 8 and using multiplication modulo 20.

Take the numbers {1, 3, 5, 7} under multiplication modulo 8. They too have the property that the product of any two of them is again one of them. Now multiply each number by 3 and use multiplication modulo 24, or multiply each by 5 and use multiplication modulo 40. In each case which number is the identity, and why? Generalize. Why doesn't it work to multiply by 2 and use multiplication modulo 16?

Suggestions

➤ Perhaps there is a relationship between the multiplier (the 6) and the modulus with respect to which the remainders are being taken which allows you to predict which element will be the identity in general.

What is interesting is not so much predicting which number will be the identity, but seeing when and why the construction works. These 'odd' looking groups are useful as examples for students who otherwise may believe that which element is the identity is 'obvious'.

Power Groups

Given a finite group G, let $P(G)$ denote the power set of G, that is, all subsets of G. Put an operation on subsets by $A \circ B = \{ab: a$ in A and b in $B\}$. Which collections of subsets of $P(G)$ form groups under this operation?

Suggestions

➤ Trying specific groups may not be all that helpful, after initially coming to understand the problem. Instead, consider what it would mean for a set to be an identity in its new group.

➤ What about inverses? Concepts of co-sets and normal subgroups may be useful.

Cubic Groups

Given any two points on a cubic, the chord through them intersects the cubic in a unique third point. This induces an operation on the real line: $x \circ y = z$ where $(z, f(z))$ is the third point of intersection of the chord through $(x, f(x))$ and $(y, f(y))$. If x and y coincide, then the chord is taken to be the tangent at that point. Is this operation well defined? (In other words, is it always possible to find one and only one value for $x \circ y$?) Is the operation associative? Does this operation have an identity? Is it commutative? Is there an inverse for each real number?

Suggestions

➤ It might be helpful to start with a really simple cubic, or perhaps with representatives of the three possible kinds of cubics.

➤ The construction extends to twisted cubics in space, that is, curves of the form (t, t^2, t^3): t in R.

➤ More generally, for a polynomial of degree d, the polynomial of degree $d - 2$ through any $d - 1$ points on the curve must intersect the curve in a dth point, so an analogous operation can be defined sending $d - 1$ tuples of reals to R. What sort of mathematical structure do these exemplify?

j-to-k Functions

A function f is said to be *j-to-k* if the number of values in the domain which map to any distinct set of k values in the co-domain is at most j. In other words, for any set S in the domain, if the cardinality of $\{f(s): s \in S\}$ is greater than k, then the cardinality of S is greater than j. A function f is said to be strictly *j-to-k* if every set of cardinality j in the domain has an image set has cardinality k.

Suppose the composite function $f \circ g$ is *j-to-k*. What can be deduced about f and g? What if it is strictly *j-to-k*?

Suggestions

➤ Try starting with $k = 1$ and allowing j to vary.

➤ The case of $j = k = 1$ is likely to be familiar. Trying to extend the familiar often reveals new insight.

Properties Preserved by Conjugation

Let f be a one-to-one function so that its inverse is also a function. The function $f^{-1} \circ g \circ f$ is the conjugate of g by f. What properties (such as being one-to-one, or onto, or continuous, or periodic etc.) of a function g are preserved by conjugation. What additional properties of f will preserve different properties of g?

Suggestions

➤ Care is needed in trying examples because, if they are too simple, they may give misleading results.

➤ Taking an action or operation on a mathematical object (here, conjugation of a function) and seeing what properties carry across and which do not provides insight into the role of those properties.

Perimeter, area and volume

The concepts of perimeter, area and volume are of practical importance. The concepts are met in elementary grades but there are subtleties of definition that can engage students of mathematics at the highest levels. The formulae range from the elementary to advanced.

Questions from earlier chapters

Tethered Goat	Chapter 2	Interpreting a verbal description into a diagram and then using geometric relationships to identify the component shapes of the area.
Tethered Goat Silo	Chapter 10	Using calculus techniques to find area.

Additional questions

Area and Perimeter

- What actions can be performed on a shape and yet leave the perimeter invariant?
- What actions can be performed on a shape and yet leave the area invariant?
- What actions can be performed on a shape and leave both area and perimeter invariant?

Suggestions

➤ Try restricting attention to a specific class of shapes, such as rectilinear (only right-angles) in order to get a sense of the possibilities. Then try to extend or modify the description of those actions so as to apply to polygons more generally, or indeed to all non-self-intersecting shapes.

One of the contributions of twentieth-century mathematics has been to add the perspective that focusing on actions and the properties they leave invariant often leads to powerful definitions in terms of applications outside of mathematics as well as inside mathematics. Pure mathematics often then studies the actions themselves as objects, and considers invariant actions upon them!

Extending Area

Is there a sensible way to define the area of a self-intersecting polygon?

Suggestions

➤ 'Sensible' here means that it fits with the ordinary definition when applied to non-self-intersecting polygons, and that it fits with commonsense properties of area. Is there an important difference between defining the sensible area of a concave polygon and a self-intersecting polygon?

It is only by trying to extend ideas that you begin to appreciate the constraints that make them useful. If you ask a dynamic geometry program for the area of a self-intersecting polygon it may give you zero!

Archimedes' Regions

The shaded region in the first figure is called an *arbelos* (because it looks like a tailor's knife of that name), and the shaded region in the second figure is a *salinon* (because it looks like a Greek salt cellar). Both are made from semicircles. Find the areas of the shaded regions in terms of the distances marked *h*.

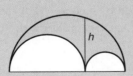

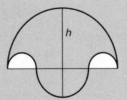

Suggestions

➤ Denote things you don't know by letters, so that you can express geometrical relationships algebraically.
➤ Keep track of what you WANT and what you KNOW.

Ratio Sum and Product

Let *P* be any point inside a triangle. Draw in the lines *APX*, *BPY* and *CPZ* as shown.

Find $\dfrac{PX}{AX} + \dfrac{PY}{BY} + \dfrac{PZ}{CZ}$.   Find $\dfrac{AZ}{ZB} + \dfrac{AY}{YC} - \dfrac{AP}{PX}$.   Find $\dfrac{AY}{YC} \times \dfrac{CX}{XB} \times \dfrac{BZ}{ZA}$.

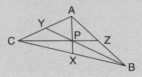

Suggestions

➤ Think about using the ratios to calculate ratios of areas. Try to relate all your areas to the area of *ABC*. If two fractions are equal, then they are also equal to the fraction whose numerator is the sum of their numerators and denominator is the sum of their denominators.

➤ Using dynamic geometry can only verify via measurement.

➤ Thinking to switch from ratios of lengths to ratios of areas might be triggered by memory of exploiting the fact that triangles with bases on the same line and a common vertex not on that line have related areas.

➤ What happens if *P* is outside the triangle? The product is known as Ceva's theorem.

Ice Cream Cone

A sphere of ice cream is inserted in and held by a cone. What radius of the sphere will have the maximum amount of ice cream inside the cone for a given cone angle?

Idea from Matt Richley in Jackson and Ramsay (1993) p. 119.

Suggestions

➤ This is likely to require more work than it seems at first. First work on subquestions such as the volume of the partial sphere. It may help to seek ways to express quantities in terms of parameters that make it relatively easy to do calculations.

Area Bisectors

Characterize those planar shapes with the property that there is a fixed point *P* which lies on every straight line that bisects the area of the shape. Change area to perimeter.

Suggestions
➤ Watch out for unjustified assumptions about centres-of-gravity and area bisection!

Geometrical reasoning

The questions in this section draw on general geometrical and spatial reasoning, and are mostly accessible without extensive knowledge of geometry. Geometric reasoning is supported by explicit use of visualization and mental imagery, as well as systematic thinking and analysis. It is assisted by observation of the physical world with attention guided by learned mathematical constructs, such as lines, planes and intersections.

Questions from earlier chapters

Envelopes	Chapter 2	Using geometrical relationships.
Cubes Cubed	Chapter 2	Using spatial visualization of a cube made up of smaller cubes.
Painted Tyres	Chapter 4	Reasoning based on the structure of a bicycle and the circumference of a circle.
Quad-cut Triangles	Chapter 5	Using geometrical constraints; trial and error may make some progress, but reasoning systematically about possible cases using geometrical relationships is more convincing.
Multi-facets	Chapter 6	Using geometrical relationships in three dimensions.
Paper Knot	Chapter 6	Geometrical relationships emerging from folding paper.
Seesaw	Chapter 8	Expressing geometrical constraints.
Paper Bands	Chapter 8	Questions arising from paper folding, using spatial relationships.
Cube Painting	Chapter 10	Using spatial relationships.
Right Angles	Chapter 10	Clarifying definitions; looking for relationships imposed by geometrical constraints.
Shadows	Chapter 10	Expressing common experience mathematically.

Additional questions

Geometrical Iterations (1)

Draw a triangle ABC. Take any point P_0 in the plane of the triangle. Reflect P_0 in line AB to give P_1; reflect P_1 in BC to give P_2 and reflect P_2 in CA to give P_3. Repeat the cycle once more to get P_6. Now consider the segment P_0P_6 as P_0 varies. Explain.

Suggestions
➤ Dynamic geometry software makes formulating a conjecture much easier, but why must it happen?

Geometrical Iterations (2)

Choose three lines L_0, L_1 and L_2 in a plane (no two are parallel). Put a point P_0 on a circle in that plane. At stage k, draw a line perpendicular to line L_k (mod 3) through point P_k. Let P_{k+1} be the other intersection of the perpendicular and the circle. Keep going. Explain why the phenomenon happens!

Again choose three lines L_0, L_1 and L_2 in the plane (no two are parallel). Put a point A_0 on L_0. Draw a perpendicular to L_0 through A_0 and label where it meets L_1 as B_0. Let the perpendicular to L_1 through B_0 meet L_2 at C_0, and the perpendicular to L_3 through C_0 meet L_0 at A_1. Keep going. Explain why the phenomenon happens!

Suggestions
➤ Dynamic geometry software is very convincing, but why must it happen?
➤ Try changing the number of lines. Does taking 'simpler' triangles help?

Shape Assembly

Imagine a triangle. Become aware of the choices you have in selecting your triangle.

Now imagine a second copy of your triangle. How many different quadrilaterals can you make by gluing your two triangles together edge to edge? Try to work mentally as long as possible!

● *Focus on edges*: what edge lengths do your quadrilaterals have?
● *Focus on angles*: what angles do your quadrilaterals have?

Which of your quadrilaterals can be formed in two different ways from different pairs of triangles? From the same pair of triangles?

Suggestions

➤ Interesting variants include making shapes involving three, four or more copies of a given triangle (especially an equilateral triangle), squares, rectangles of a given size, congruent quadrilaterals, regular hexagons and even cubes or congruent cuboids. Counting how many different ones can be made is actually very challenging beyond the first five or six, but provides excellent practice in recognizing relationships and detecting congruence of awkward shapes.

➤ The more you push yourself to stay in the world of mental imagery, the better you will become at working mentally. Then, when you draw a diagram, you can use it more powerfully to supplement and stabilize your mental imagery.

Shape Partitioning

What shapes have the property that they can be cut along a single straight line to form two congruent shapes similar to the original? Two similar shapes similar to the original? Two similar shapes?

Suggestions

➤ Try building up shapes. The aim is to classify all possible such shapes. How many sides can they have?

➤ What about two cuts to form three similar shapes?

➤ What about three dimensions?

➤ Having found a few such shapes, trying to show that you have found them all involves logical and geometrical reasoning.

Rectilinear Winding Numbers

Draw a closed rectilinear path (all changes of direction are right angles; eventually the path joins up to the starting point), allowing edges to cross each other. Find a relationship between the number of clockwise right angles, anticlockwise right angles, and the number of complete revolutions as you traverse your polygon in one direction. Is that relationship a property for all rectilinear polygons?

What if the right angles are changed to $\pm\theta$? For which θ can such a polygonal path close up?

Suggestions

➤ This is a development of *Right Angles* in Chapter 10.

➤ Taking a core idea (here, right angles) and exploring what can be constructed with it provides learners with opportunities to rehearse the vocabulary and the techniques while finding out things for themselves, especially recognizing and expressing relationships.

Rational Geometry

A point is said to be *rational* if both of its coordinates are rational numbers. A straight line is said to be *rational* if there are two rational points on it. A circle is said to be rational if there are three rational points on it.

- Can a line have just one rational point on it? Can a circle have just one or just two rational points on it?
- Can a rational line have more than two rational points? Can a rational circle have more than three rational points?

Suggestions

➤ Constructing your own examples can include looking for the simplest example as well as the most general, in order to be sure that all possible cases have been considered.

Reasoning

In these questions the focus is on logical reasoning rather than on any specific algorithm or concept. In some of these questions, working systematically through multiple possibilities is the key; others require being alert to hidden possibilities and unwarranted assumptions. Others involve classic patterns of reasoning, such as *reductio ad absurdum*.

Questions from earlier chapters

Ladies Luncheon	Chapter 2	Reasoning systematically about possibilities in logical relationships.
Quick and Toasty	Chapter 2	Reasoning about properties of material objects; offers introductory experience of optimization.
Painted Tyres	Chapter 4	Reasoning based on the structure of a bicycle and the circumference of circle.
Furniture	Chapter 4	Reasoning based on the spatial properties of a grid.
More Furniture	Chapter 10	
Eureka Sequences	Chapter 5	Reasoning making use of counter-examples.
Fifteen	Chapter 5	Finding a way to present relationships so as to aid reasoning.

(continued)

Milkcrate	Chapter 5	Reasoning based on the spatial properties of a grid.
Nine Dots	Chapter 6	Reasoning based on overcoming self-imposed but unnecessary constraints.
True or False	Chapter 6	Reasoning based on logic.
Cartesian Chase	Chapter 10	Reasoning in a game context based on spatial relationships: even and odd numbers play a role because it is a two-player game.
Fare is Fair	Chapter 10	Reasoning about spatial relationships.
Flipping Cups	Chapter 10	Reasoning based on locating invariants.
Glaeser's Dominoes	Chapter 10	Reasoning based on spatial relationships.
Nullarbor Plain	Chapter 10	Reasoning based on spatial directions.
Thirty-one	Chapter 10	Reasoning about strategy in a game context based on simple arithmetic.

Additional questions

See also *Moving Means* and *Remainders of the Day*.

Birthday Greetings

I was once sent a card saying 'Happy
$$\begin{bmatrix} 16 & 15 & 21 & 12 & 18 \\ 5 & 4 & 10 & 1 & 7 \\ 20 & 19 & 25 & 16 & 22 \\ 6 & 5 & 11 & 2 & 8 \\ 11 & 10 & 16 & 7 & 13 \end{bmatrix}$$
 'th Birthday'.

The instructions said to select five numbers, one from each row and each column, and take their sum. How does it work? Make up your own.

Suggestions
➤ What actions can you perform on the array without it changing the sum? Seeking actions that leave a property invariant is a powerful technique for simplifying a situation.

Same Birthdays

Suppose that in a particular crowd, among any five distinct people, at least two have the same birthday.

What is the least number of people for which, among any group of that size, it must be the case that at least five have the same birthday. What other assertions like this can be made?

Generalize!

Suggestions

➤ It might be possible to establish a bound on the number of distinct birthdays present amongst the crowd. It might also be possible to work out a worst-case or largest group of people with no five having the same birthday.

Paying attention to how you work out the particular case enables you to generalize.

Rectangular Max–Min

Draw up a rectangular grid and put a number in each cell.

● For each row, circle the maximum in that row and note the minimum of these.

● For each column, put a square around the minimum and then note the maximum of these.

Is there anything that can be said in comparing the minimum of the row-maxima and the maximum of the column-minima?

Suggestions

➤ Try specializing, choosing an example that is not too special. Use it to try to see what is going on structurally, for example, by 'specializing' your attention to a particular row and a particular column.

➤ Try replacing max by sum; try replacing min by sum; try replacing sum by product;

➤ Try replacing max by arithmetic mean and min by geometric mean (why does this change the reasoning?), or max by arithmetic mean and min by harmonic mean.

➤ Why does it not work to replace max by arithmetic mean but to leave min?

➤ What happens if you use arithmetic means for both rows and columns?

➤ This is an ideal opportunity to specialize to a really simple case to find out what if anything must be the case, and then to Watch What You Do as you work on a more complicated case in order to try to see what is going on. In order to make contact with the underlying structure, you need to do more than simply calculate the maximum in each row, etc. You need to organize things so that you can see relationships, particularly the maxima and the minimum of these.

➤ Specializing can mean not just choosing a particular array of numbers, but also focusing on some particular part of a more complex configuration. For example, try focusing on the maximum of a particular row and the minimum of a particular column. What can you say about their relationship?

➤ In order to probe what is really going on structurally, it helps to change the max and min to other 'statistics' (representative values) in order to see what works and what does not.

See also *Generalized Max–Min*.

Generalized Max–Min

S and T are two families of sets of real numbers with the property that each pair of sets $\{s, t\}$ with s in S and t in T have at least one element in common and with the property that every set has a least upper bound and a greatest lower bound.

Now for each s in S, let s_M be the least upper bound of the numbers in s, and let σ denote the greatest lower bound of these. Similarly, for each t in T, let t_m be the greatest lower bound of the numbers in t, and let τ denote the least upper bound of these.

Compare σ and τ.

Construct counter-examples for removing the constraint that each pair of sets, one from each family, has a non-empty intersection.

Suggestions

➤ Pay attention to your reasoning in the finite case (see *Rectangular Max–Min*) and see if it can be extended to the infinite case. Constructing counter-examples when a condition is removed is good mathematical practice.

References

Jackson, M. and Ramsay, J. (eds) (1993) *Questions for Student Investigation*. MAA Notes 30. Washington: Mathematics Association of America, p. 119.

Maclaurin, C. (1725) *An Introduction to Mathematicks*. Unpublished ms 2651. Edinburgh: Edinburgh University, p. 37.

Rabinowitz, S. (ed.) (1992) *Index to Mathematical Problems 1980–1984*. Westford: Math-Pro Press.

Ransom, W. (1952) E1021. *Mathematical Monthly* **59**(6), 407.

Ransom, W. and Braun, J. (1953) E1021. *Mathematical Monthly* **60**(2), 118–19.

Sibley, T. (2008) Sublimital analysis. *Mathematics Magazine,* **81**(5), 369–73.

Watson, A. and Mason, J. (2006) *Mathematics as a Constructive Activity: Learners Generating Examples*. Mahwah: Lawrence Erlbaum.

12

Powers, themes, worlds and attention

This chapter is an extended glossary concerning core elements of mathematical thinking as developed in the book. Some of the constructs described here have come to articulation since the original edition was published so, although they are not illustrated and identified by explicit commentary on problems in the core of the book, they are nevertheless useful in developing your own awareness of your mathematical thinking to help you be better placed to help others develop their own mathematical thinking.

Natural powers and processes

Our claim is that there are natural powers possessed by every child, and that thinking mathematically is really about learning to use these powers in mathematical ways and in the exploration of mathematical problems. These powers are natural in the sense that they are integral to human intelligence and used across fields of human activity. However, for most students, learning most mathematics is far from a natural endeavour, even though there are roots in common experience. In the sense of Lev Vygotsky, learning mathematics is a 'scientific' endeavour and most people need to be in the presence of a teacher, someone more experienced, at least some of the time, in order to understand how processes fundamental to human intelligence are applied in this domain.

Specializing and generalizing

George Pólya used the term *specializing* where he might have used the less pleasant sounding term *particularizing*. Specializing means considering a simpler case (fewer dimensions, fewer variables, fewer parameters, simpler numbers) or a special case (when some numbers are zero or one or some other value that reduces the complexity). However, what is often missed by students is that the act of specializing is not to get an answer as such, but rather to watch what you are doing when you examine a particular instance or case, in order to recognize relationships that might generalize to all other cases and instances. In

other words, the purpose of specializing is to become aware of structural relationships in order to generalize. As summarized in Chapter 1, specializing can
be done:

- randomly to get the feel of the questions;
- systematically to prepare the ground for generalizing;
- artfully to test the generalization.

Generalizing is the process of 'seeing through the particular' by not dwelling in
the particularities but rather stressing relationships. Caleb Gattegno noticed
that whenever we stress some features we consequently ignore others, and this
is how generalizing comes about. Sometimes it is useful to distinguish between
two kinds of generalizing: empirical and structural. *Empirical generalization*
comes about when you look at several, sometimes many, cases or instances and
ask yourself what is the same about them all. By stressing the sameness (and
consequently ignoring differences), you effectively generalize. When you articulate the sameness you produce a conjectured general property, which then has
to be justified by reference to structure. *Structural generalization* arises when
you recognize relationships in one or very few cases. Perceiving these relationships as properties, your articulation is again a conjectured generalization which
then has to be justified by reference to underlying structure. As summarized in
Chapter 1, and elaborated in subsequent chapters, generalizing means detecting a pattern leading to:

- what seems likely to be true (a conjecture);
- why is it likely to be true (a justification); and
- where it is likely to be true, that is a more general setting of the questions
  (another question!).

The difference between scientific induction and empirical generalization is that
in science there is no way to be certain your conjectured general properties are
correct. Nature never gives a yes or no answer. Empirical generalization, the
process of finding a conjecture about what might be true from numerous
instances, is similar to the process of scientific induction. However, by contrast,
in mathematics structural generalization is possible and you can go further and
justify conjectures by logical reasoning on the basis of agreed properties. Note
that *mathematical induction* is different again, being a form of reasoning about
conjectures concerning a sequence of relationships, usually associated with the
natural numbers.

As is evident throughout the book, generalizing and specializing go hand in
hand. Their relationship is captured in the slogans

- seeing the particular in the general;
- seeing the general through the particular.

Whenever a mathematical problem is solved, and whenever an example of a mathematical concept is encountered, it is useful to ask yourself what the *dimensions of possible change* might be. This language follows that of Ference Marton who suggested that learning is about discerning the dimensions of possible change under which an example remains an example. Thus, given a drawing of an angle, what can be changed and still that figure presents the same angle? Actions that do not affect the angle include changing the lengths of the arms, or translating or rotating the diagram in space. These actions all preserve the angle, and so are dimensions of possible variation. You cannot be said to appreciate and understand a concept if you are not aware of dimensions of possible variation, or, put another way, whenever you become aware of a further dimension of possible change, your appreciation of the concept deepens.

When some attribute or feature can be changed, it is important to consider the *range of permissible change*: for example, the arms of an angle must have positive length; a problem involving counting whole objects does not permit fractional values. The formula $2^n - 1$ for the number of creases in *Paper Strip* (Chapter 1 and elsewhere) can be evaluated when n is not a natural number, but does not appear to have any sensible meaning in the context of the problem. In other problem contexts involving similar formulae (see, for example, *Exponential Percentages*, Chapter 11) values of n other than integers are meaningful. The adjectives *possible* and *permissible* are used because it is often the case that a teacher is aware of possible features that can change, whereas a learner may not be aware of all of them, so when *possible* comes to mind, it is a reminder to check that the audience is aware of the pertinent dimensions. Similarly, even when learners are aware of something that can change, they may not be aware of the full extent of that change. Indeed, the way mathematicians use the word *number* involves several extensions from the original counting numbers. An audience may not immediately be aware of the full range of permissible change being considered, and may confine themselves to a more restricted kind of number. A good example of this arose in *Consecutive Sums* (Chapter 4) when solving a problem about sums of positive numbers was assisted by extending sums to include negative numbers.

Conjecturing and convincing

In a mathematically productive atmosphere, everything that is said is treated as a conjecture. Instead of letting possibilities tumble around in the mind like clothes in a tumble drier, getting more and more confused, it can often help to bring a conjecture to articulation so that it can be looked at dispassionately. Pólya used to say that once you have made a conjecture, you must not believe it, but rather try to see how it needs to be modified.

Once a conjecture has been made, the aim shifts to trying to justify it mathematically. Since the root meaning of the word *theorem* has to do with 'seeing', a conjecture can be taken as a way of seeing a situation, and a mathematical proof consists of reasoning that convinces others that they too can see what you are 'saying' and 'seeing'. Developing mathematical reasoning involves trying first to convince yourself, then a friend who asks pointed but friendly questions, and then a sceptic or 'enemy' who refuses to take anything at face value and needs to be convinced through mathematical reasoning.

Whereas accumulating some examples (particular or special cases) can support an intuitive 'sense' or conjecture, ultimately what is needed is a sequence of statements calling upon already agreed properties, which follow logically one from another. These ideas have been elaborated in Chapters 4 to 7.

Imagining and expressing

By imagining we include all forms of mental imagery, not only images like 'pictures in the mind', but also any recalled sense-based experience. Neither this power nor the power to express in different forms what is being imagined were explicitly mentioned as processes in the first edition, but they are fundamental to thinking of any type, and to thinking mathematically in particular. To anticipate, to have an expectation, is to draw upon imagination; to articulate some relationships and to propose them as properties that hold in many cases involves mental imagery. So every time you plan or prepare, you are using mental imagery; every time you consider a possibility, you use mental imagery; every time you become aware that you have recognized a mathematical relationship you use mental imagery.

Imagery alone is solipsistic. Learning to express what you are imagining, to capture and 'tie down' distinctions and relationships, to articulate or express perceived properties is to bring imagery to expression. It is possible to use material objects, diagrams and pictures, voice tones and gestures, words and symbols to express discerned objects, recognized relationships and perceived properties. Learning that what is experienced inwardly is not shared with others until it is expressed in a form to which others can relate is a contribution which thinking mathematically can make to the general social development of learners.

Whenever you get stuck, it can help to find someone to talk to about the sticking point. Articulating is a way of becoming aware of previously subconscious stressing and ignoring, and so sometimes offers directions to pursue that were previously overlooked.

Stressing and ignoring; extending and restricting

Gattegno pointed out that human beings naturally stress some aspects of an object, and consequently ignore others. For example, looking at the numeral 347 you might notice the relation that $3 + 4 = 7$ and, if so, this opens up a space of three-digit base-10 numerals for which the sum of the first two digits is the third, or perhaps the sum of two of the digits is the third. A relationship recognized between the digits becomes a property which other numerals may or may not possess. It is through stressing some features and consequently ignoring others that generalization comes about and relationships become properties. It is a mathematical generalization when the relationship being turned into a property is mathematical.

Sometimes it is important to stress, and sometimes to ignore: stressing the meaning of a variable may not be helpful when trying to solve an equation it is in; stressing the workings of addition algorithms is pertinent to *Palindromes* (Chapter 1) but irrelevant to *Paper Strips* (Chapter 1). Learners are often trapped when they have to pay attention to processes (e.g. the details of how to solve equations) when they are learning about a new concept that requires them.

In mathematics, the action of either extending or restricting meaning is a manifestation of amplifying and diminishing, of stressing and ignoring. For example, instead of considering primes in the context of all numbers, consider primes in the restricted system of numbers congruent to 1 modulo 3 under multiplication, or extend to the system of numbers of the form $a + b\sqrt{d}$ for some fixed d, and integers a and b (see *Remainder Primes* in Chapter 11). In both cases, changing the domain of what counts as a 'number' in the context of primes sheds light on the nature of primes and the role they play in arithmetic.

Classifying and characterizing

To classify things is perfectly natural. That is in effect what language does for us. Nouns and verbs are general, so when we use one of them we classify that which we are thinking about as belonging to, or having the required properties associated with, that word. Of course natural language has very fuzzy boundaries, so that an object in one context might be classified quite differently in another. For example, a tree stump acts as a chair when camping, but not at a formal reception; a piece of plastic in the shape of a triangle 'is' a triangle in one context, but a triangular prism in another. A house number carries ordinal number properties of succession in a sequence, but being a perfect square, cube or prime is not relevant.

To classify something is thus to perceive it as an instance of a property, having first discerned 'it' from its surroundings. To characterize it is to produce an alternative collection of properties so that anything belonging to the classification satisfies those properties, and anything satisfying those properties belongs to the classification. A pervasive mathematical theme is to classify objects by properties, and then to characterize those properties by means of other properties. Thus an even whole number has the defining property of being exactly divisible by 2; it is also characterized by whether it ends in 0, 2, 4, 6 or 8 when expressed in base-10; the property of a number leaving a remainder of 1 when divided by 3 is also characterised by the number being of the form 1 more than a multiple of 3. The latter characterization is easier to extend to negative numbers than the remainder version; indeed it provides a consistent extension of remainders to negative numbers. In *Flipping Cups* the challenge is to characterize possible configurations without having to test all the possibilities, by finding a condition on the cups.

Classifying and characterizing as natural powers arise frequently in concert with the mathematical theme of *doing and undoing* (see later).

Overview

These powers have been demonstrated by any child who can talk, because the acquisition of language already calls upon all of them. The question to be asked is whether in lessons learners are being encouraged to use, develop and become aware of their own powers, or whether the textbook and the teacher try to do the work for learners, thereby blocking the learners from thinking mathematically for themselves. Learning to act within a discipline such as mathematics, and indeed in different sub-disciplines of mathematics, involves learning to use these powers in domain-specific ways.

Mathematical themes

Doing and undoing

Whenever you find that you can perform a mathematical action or resolve a mathematical question (a 'doing') further explorations are available by reversing the action and asking 'undoing' type questions. For example, if you can solve a problem, ask yourself what other similar problems would give the same result, and what possible results like that are possible for similar questions. You can go further by exploring what happens if you interchange what is wanted and what is given. Very often the new problem involves creativity. For example:

- if the doing-action is 'multiplying', then the undoing is factoring, which often has multiple answers, and leads to the notion of *primes* as the numbers or expressions which cannot be factored;
- if the doing-action is 'adding' then the undoing is 'the story of . . .' because there are usually many ways to represent a number as the sum of two other numbers. Instead of being given two numbers and being asked for the sum, you can be given one number and the sum and seek the other number (subtraction);
- if gluing triangles together along their edges to make a polygon is taken as the doing-action, then decomposing polygons into triangles is the undoing.

It can be done many ways and there are difficult things to prove such as that any non-self-intersecting polygon can be decomposed into a triangle and a non-self-intersecting polygon with fewer vertices;

- if gluing polyhedra together along congruent faces is the doing-action, then decomposing polyhedra by a plane through vertices and edges is the undoing. Prime polyhedra are those for which there is no such decomposition;
- if the doing-action is solving a pair of linear equations, the undoing is finding all pairs of linear equations having that same solution.

In Chapter 11, there are many examples of how asking undoing type questions turns procedural mathematics into rich mathematical investigations.

Invariance in the midst of change

Many of the theorems of mathematics can be seen as stating something that remains invariant under some other permissible change. For example:

- adding the same thing to two numbers leaves their difference invariant; multiplying two numbers by the same non-zero number leaves their ratio invariant;
- two fractions are equivalent (their value as rational numbers remains invariant) when the numerator and denominator are multiplied by the same number;
- the sum of the angles of any planar triangle is invariantly 180 degrees, no matter how the triangle changes;
- the area, angles and lengths of a polygon remain invariant under translation, rotation and reflection;
- the area of a triangle remains invariant when one vertex is moved along a line parallel to the opposite edge;
- the angle between two straight lines is invariant under translation by either of the lines (the basis for theorems about angles related to parallel lines and the definition of translation!).

In any mathematical situation it can be informative to ask yourself what actions can be performed and still the relationship of interest remains invariant. For example, in both *Rectangular Max–Min* and in *Birthday Greetings* it really helps to make use of actions that leave the problem invariant but which organize the rows and columns usefully. In *Arithmagons* it helps to find the 'invariant' sum of the entries from which everything else can be deduced.

Freedom and constraint

Pólya distinguished two types of problem: *problems to find* and *problems to prove*. Any problem 'to find' can be thought of as a construction task: to construct all objects that satisfy the constraints given in the problem. By starting with no constraints you can consider the freedom of choice available. As each constraint is added, the freedom is restricted. By seeking the most general solutions at each stage you can build up a solution to the original problem through constraining solutions to fewer constraints. Sometimes this can be really helpful.

For example, in *Consecutive Sums*, allowing yourself the freedom to use negative numbers in the sum provides access to an underlying structure connected with odd divisors of the number. In *Nine Dots*, allowing yourself the freedom to break an assumed constraint makes a resolution possible.

Mathematical worlds

Thinking mathematically involves movements between different worlds of experience. Based on insights of Jerome Bruner which can be traced back to ancient psychologies of the Indian subcontinent, it proves useful to think in terms of:

- a world of confidently manipulable 'objects' which may be material objects in the physical world, but may be images and symbols. The point is that when complexity threatens to overwhelm, it is perfectly natural and sensible to retreat to more confident ground. This is precisely what specializing achieves;
- a world of intuitions and 'senses of', of mental imagery in all its richness, usually pre-articulated or at best barely expressible;
- a world of abstract symbols and signs that are not yet immediately and confidently manipulable. Once they become confidently manipulable, they slip into the first world!

These ideas are touched on in Chapter 9.

Study of mathematics, or perhaps of any concept-ridden domain, involves gaining sufficient familiarity and facility with constructs to be able to use them to express more precisely and clearly what you are thinking, what you are attending to and how you are attending to it. As familiarity grows, ideas and notions become firmly established concepts, and progressively become part of how you perceive, conceive and experience. Abstract symbols and signs become concretely manipulable, as if they were themselves 'concrete'. So it is movement between these worlds which is involved in developing an understanding, becoming appreciative, getting to know.

These three worlds provide a background structure for the process of building mathematical models of situations arising in the material or in the mathematical world: perceiving a situation in mathematical terms through recognizing relationships and conceiving of them as properties that can hold in many situations, and expressing these properties in some form, usually but not always algebraic. Thus a problem starts in a situation which is reasonably familiar or specific; through the use of mental imagery relevant features are discerned and identified, and possible pertinent relationships are recognized and expressed, becoming properties in the process. When these relationships are expressed mathematically, you enter the mathematical worlds of symbols, and through manipulating them you reach for a mathematical resolution. This is then checked out through the world of mental imagery, and then back in the original setting, to make sure that pertinent assumptions are explicit and reasonable, and that the resolution goes some way to resolving the original problem.

Pedagogically, teachers often use structured relationships as expressed in some familiar material world situation, as a model for a mathematical concept: cubes, flats and rods for base-10 notation, balances for linear equations, and empty number-lines for working with numbers are common. But these models are only effective when there is complete familiarity and when students recognize what it is about the model that is appropriately carried across to the mathematical concept.

Multi-base arithmetic blocks (cubes, flats and rods) model the size of a number by the volume of the representing pieces. The base-10 structure of the numeral is well represented in the collection of cubes, flats, etc., but the *place* value structure of the numeral is not represented by this material: one cube and two flats represent 1200 regardless of whether the cube is to the left or right of the flats. Balances as a model for equations run into difficulties when negatives are involved.

Attention

Certain adjectives and nouns have been used repeatedly in this chapter, in order to link all of the powers and themes to the movements of attention. This section elaborates on these and suggests that what solves problems is movements of attention.

Sometimes people look at a scene or situation, a poster, exercise or diagram, and gaze or stare at it. They take in the whole, they hold the whole as a whole. Of course they are aware in some sense of components that make up the whole, but the dominant feature of their attention is gazing. One of the reasons for gazing is to get a holistic sense, and to allow metaphoric resonance and metonymic triggers to bring to mind possible actions.

Sometimes attention is dominated by discerning details, picking out elements, stressing boundaries, singling out 'sub-wholes' for possible gazing. 'This not that' is typical of discerning details. All of learning can be seen as a form of learning to discern, to make distinctions not previously made.

Sometimes attention is concerned with recognizing relationships between discerned elements in the situation. A great deal of mathematics can be described in terms of recognizing and articulating relationships. When these turn into perceived properties being instantiated, mathematical generality becomes possible. When a relationship between discerned elements is reconstrued as an instantiation of a more general property, mathematical conjectures and associated reasoning become possible.

When reasoning is based only on previously agreed properties (rather than calling upon anything known about the particular object), then the way is open to mathematical theories. Agreed properties serve as axioms, and all other properties are deduced from these or added explicitly as fresh axioms.

These five different types or foci of attention are characteristic of mathematical exploration. They rarely happen in a single sequence; rather attention tends to move quickly between and around these states. By being aware of these states, by developing a taste for what they are like, it becomes possible to invoke them intentionally rather than be subject to the whims of habit and personal dispositions.

Summary

Seeing mathematical thinking processes as use of natural human powers leads to the question of whether learners are being encouraged to become aware of, use and develop those powers, or whether those powers are being usurped by text and teacher. Probing the use of those powers in mathematics leads to recognition of core themes that occur again and again, and which provide connections between apparently disparate topics and problems. Probing the experience of thinking mathematically leads to the question of how attention shifts, sometimes quickly, sometimes slowly. The purpose of the questions posed in this book is to provide opportunity for exploring your own experience so as to be sensitized to the experiences of others.

Bibliography

Adams, J. (1974) *Conceptual Blockbusting*. San Francisco: Freeman.

Banwell, C., Saunders, K. and Tahta, D. (1986) *Starting Points: For Teaching Mathematics in Middle and Secondary Schools*, updated edn. Diss: Tarquin.

Conway, J. and Guy, R. (1996) *The Book of Numbers*. New York: Copernicus, Springer-Verlag.

Dudeney, H. (1958) *Amusements in Mathematics*. New York: Dover.

Gattegno, C. (1963) *For the Teaching of Mathematics*. New York: Educational Explorers.

Hofstader, D. (1979) *Gödel, Escher, Bach: an eternal golden braid*. London: Harvester.

Jaworski, J., Mason, J. and Slomson, A. (1975) *Chez Angelique: The Late Night Problem Book*. Milton Keynes: Chez Angelique Publications.

Maclaurin, C. (1725) *An Introduction to Mathematicks*. Unpublished ms. 2651, Edinburgh University p. 37.

Moessner, A. (1952) Ein Bemerkung über die Potenzen der natürlichen Zahlen. S.–B. Math.-Nat. Kl. Bayer. Akad. Wiss., **29**(14), 353b.

Noelting, G. (1980) The development of proportional reasoning and the ratio concept part I: differentiation of stages. *Educational Studies in Mathematics*, **11**(2), 217–53.

Rabinowitz, S.(ed.) (1992) *Index to Mathematical Problems 1980–1984*. Westford: MathPro Press.

Sibley, T. (2008) Sublimital analysis. *Mathematics Magazine*, **81**(5), 369–73.

Streefland, L. (1991) *Fractions in Realistic Mathematics Education: A Paradigm of Developmental Research*. Dordrecht: Kluwer.

Wason, P. and Johnson-Laird, P. (1972) *Psychology of Reasoning: Structure and Content*. London: Batsford.

Watson, A. and Mason, J. (2006) *Mathematics as a Constructive Activity: Learners Generating Examples*. Mahwah: Lawrence Erlbaum.

We were particularly strongly influenced by:

Bennett, J.G. (1969) *Creative Thinking*. London: Coombe Springs Press.

_____ (1978) *Deeper Man*. London: Turnstone.

Bloor, D. (1976) *Knowledge and Social Imagery*. London: Routledge and Kegan Paul.

Bruner, J. (1956) *A Study of Thinking*. New York: Wiley.

Bruner, J. (1966) *Towards a Theory of Instruction*. Harvard University Press.

Edwards, B. (1981) *Drawing on the Right Side of the Brain*. London: Stewart Press.

Gattegno, C. (1963) *For the Teaching of Mathematics.* New York: Educational Explorers Ltd.

———— (1970) *What We Owe Children: the subordination of teaching to learning.* London: Routledge and Kegan Paul.

Hadamard, J. (1954) *The Psychology of Invention in the Mathematical Field.* New York: Dover.

Honsberger, R. (1970) *Ingenuity in Mathematics.* New York: Random House.

Jackson, M. and Ramsay, J. (eds) (1993) *Questions for Student Investigation.* MAA Notes 30, Washington: Mathematics Association of America.

Krige, J. (1980) *Science, Revolution and Discontinuity.* London: Harvester.

Lakatos, I. (1977) *Proofs and Refutations: The Logic of Mathematical Discovery.* Cambridge: Cambridge University Press.

Schoenfeld, A. (1985) *Mathematical Problem Solving.* New York: Academic Press.

Polanyi, M. (1958) *Personal Knowledge.* Chicago: Chicago University Press.

Pólya, G. 1957. *How to Solve It.* Princeton: Oxford University Press.

———— (1966) *Mathematical Discovery.* (Volume I) New York: Wiley.

———— (1968) *Mathematical Discovery.* (Volume II) New York: Wiley.

Walter, M. and Brown, S. (1983) *The Art of Problem Posing.* Philadelphia: Franklin Press.

Since the publication of the first edition of this book, the authors have published several works on the same theme, among them:

Burton, L. (1984) *Thinking Things Through.* Oxford: Blackwell.

Mason, J. (1988) *Actions Into Words,* Project Update. Milton Keynes: Open University.

———— (1988) *Doing and Undoing,* Project Update. Milton Keynes: Open University.

———— (1998) *Learning and Doing Mathematics* (2nd rev. edn), York: Tarquin Publications.

———— (2002) *Researching Your Own Practice: the discipline of noticing.* London: Routledge Falmer.

Stacey, K. and Groves, S. (2004) *Strategies for Problem Solving,* 2nd edn, VICTRACC Ltd: Victoria, Australia.

Subject index

Index of questions

2 80, 81, 110, 199